Opening the East River

# Opening the East River

## *John Newton and the Blasting of Hell Gate*

THOMAS BARTHEL

McFarland & Company, Inc., Publishers

*Jefferson, North Carolina*

ISBN (print) 978-1-4766-8298-3
ISBN (ebook) 978-1-4766-4326-7

LIBRARY OF CONGRESS AND BRITISH LIBRARY
CATALOGUING DATA ARE AVAILABLE

Library of Congress Control Number 2021038628

Front cover image: *inset* John Newton in the uniform of a Major General,
(*Harper's Weekly*, Oct. 14, 1876); The Port of New York—bird's eye view from
the Battery, looking south, Currier & Ives, circa 1892 (Library of Congress)

Printed in the United States of America

*McFarland & Company, Inc., Publishers
Box 611, Jefferson, North Carolina 28640
www.mcfarlandpub.com*

For Michael and Wolf

# Table of Contents

# Preface

This book is concerned with the immense engineering achievement of General John Newton from 1866 to 1885. Assigned to clear the East River of its many obstacles, Newton was well aware that the enormous task set to him was directly tied to the economic prosperity of the nation and to its most prosperous city. New York City, thanks to its harbor, collected revenues that equaled 42 percent of the federal government's income. Almost a million dollars was collected just during the first six days of 1822, all before a federal income tax.

The book focuses on the Hell Gate section of New York's East River (which is actually a salt-water tidal strait since it contains no fresh water). In the general's time, there was a cry to rename the East River "Newton's Channel" and yet his triumph has been forgotten. It is important to see what was accomplished through many changes and many setbacks and to learn the interplay between ships, explosives, and technology and the growth of New York and the nation: 125 years after the final explosion of the nine acre Flood Rock, *National Geographic* calls Newton's design and execution "an epic moment in the history of civil engineering."[1]

The book is not a biography. It's not about the personality of the general. Of John Newton, little can be found beside his Civil War engagements and some reluctant political involvement during the War. Of his decision to fight on the Union side, he being a Virginian, little is written. There seem to be few letters, no memoirs. Typical of 19th century newspaper and periodical reporting, the press' focus in the writing was on facts, not on the person. What we can appreciate about Newton is the intelligence, the skill, and the willingness to adapt to the changes in all aspects of his work. We know his subordinates and workers were loyal, even in the most dangerous kind of work handling newly discovered nitroglycerine and dynamite.

For me, in order to understand the many changes during Newton's nearly two decades in command, it became vital to explain and analyze a wide array of topics. So, this book ties together the history of ship building, the development of explosives, the contributions of the Army Corps of

Engineers, the influence of the Erie Canal on American life, the inventions in science, the financial disasters of the period, the inventions of technological devices, the experimentation with explosives, the competition with railroads, and the dedicated inventiveness, patience and devotion to duty of General Newton who gathered all of these together.

My study for this book on the achievements of General John Newton took me back to the 1600s. The newspaper research from the later 19th century often resulted in copies of print so old that some of them were unreadable. Some reports and periodicals contained references to houses and streets that no longer exist. Likewise, many books on New York City forgo bibliography, endnotes and index. Many favor the sensational, the tawdry and the fictional.

For the reader to easily follow the process, I have divided the book chronologically, even though the official reports used the fiscal years. Thus, though the events in the annual reports covered from July 1 of one year to June 30 of the next, the reports were not published until October.

The result of more than fifteen years of research, this book includes 270 distinct references, and more than 300 endnotes showing direct use of sources.

My aim has been three-fold: first, to be as thorough as possible since the book would probably forever be the only one on the subject. Second, I have tried to keep my opinions out of the book, aiming for my work to be more like reportage than persuasion. Third, I have focused on the subject itself and on the work around it, unless the larger world around it directly affected my subject.

I have included an appendix (*A Gazetteer of the East River Obstructions*) that I hope may be useful. It is a collection of names and places, most forgotten, some still using the old Dutch designations. For example, the location for "Quinters Reef" remains a mystery to me.

down the Ohio River to the Mississippi. At New Orleans, sail east around the tip of Florida and then north in the Atlantic to New York. Traveling overland from Missouri to Oregon took nearly two years.

While progress was being made in ship design leading to greater speeds, an account of an overland ride by James S. Buckingham, titled *Travels in the Eastern and Western States of America,* was published in London in 1842. All Buckingham wanted to do was to travel from Utica in central New York to Auburn, New York, an 82 mile journey. Using stagecoaches to move to Syracuse, he was surprised to find the last leg on board a railroad that had "the cars … drawn by horses…. The rails, too, were of wood instead of iron."[3]

## New York, a Maritime City

Water was still the fastest way for commerce and generally the most comfortable way for passenger travel. New York held dominance as the premier harbor city in the third decade of the 19th century. The fact that the city ranked high among manufacturing locales only added to the need for merchants and makers to send to, and receive material from, all over the country and the world. Geography matters here too. True, the port at Boston is a voyage of 200 fewer miles closer than New York to Britain's main port, Liverpool. However, Boston is badly positioned to ship to all the ports to the south, the direction for much of the population at the time. (New York requires a sea journey of more than 300 miles.) Philadelphia is set about 125 miles from the Atlantic Ocean and lies 900 more sea miles from Liverpool than from New York Harbor.

Some quickly recognized the commercial aspects of New York's harbor. One sign of the importance of the maritime commerce in the city is the appearance of many shipbuilders. As early as 1792, Samuel Akerly built a shipyard on the East River. The Hudson and the East rivers needed ships; the ocean-going and lake vessels needed to be built. In 1800, at Cherry and Clinton streets in the city, Cheeseman and Brownne went into the trade, themselves constructing Robert Fulton's first boat.

Morrison reports in his 1909 book, the output of the city's shipyards in one year—1826—totaled "twenty-three ships, three brigs, forty-nine schooners, sixty-eight sloops, twelve steamboats, fifteen towboats and nineteen canal boats, making a total of 29,137 tons."[4] Just as ships had to be built, so ships had to be repaired. On 10th Street, the New York Dry Dock took vessels out of the water for repairs, horses supplying the pulling power. Soon, "an entire neighborhood" would be called the "Dry Dock Neighborhood," which fell between 3rd and 9th streets and between Avenue C and the East River.

So much depended on the ability of New York City to provide dockage—a wharf, a pier, a basin or a slip. By mid-century these structures to receive vessels and to store goods occupied a two-mile stretch from along the East River, down to the Battery, and up the Hudson. The areas remained so busy that by 1815, the money collected by customs grew to $14,491,739, almost 42 percent of the federal government's revenue. In seven more years, almost a million dollars was collected just during the first six days of 1822. Clearly, the port of New York City had established itself as the primary port in the country. And that port needed a navigable East River.

## *The Sandy Hook Problem*

Sandy Hook was a passageway with two major problems clearly evident. Once ships left piers ("cleared port") of the Hudson and East rivers, heading out to the Atlantic Ocean lay a space of 25 miles from the Battery. Using this way to exit New York Harbor a vessel passed first through the Upper Bay heading to The Narrows, a land width of 1.01 miles. Once past that space, a ship sailed out towards the Lower Bay. From the Lower Bay the passage to the ocean was where the problem presented itself. As the approach to the ocean neared, vessels had to contend with a wide and shallow underwater sandbar five miles long, between Sandy Hook, New Jersey, and Coney Island. Outward bound, the ship's hired pilot probably was forced to wait for high tide.

Inward bound ships from Sandy Hook laden with goods had three choices. First, a comfortable anchorage spot. Second, that berth being unavailable, in many cases a captain might hire a pilot to guide the ship through a shallow channel in the sandbar and into port. There was no choice involved. Since March 9, 1694, the law made it clear that "it is the duty of the master … to put his ship under the charge of a pilot, both on his outward and homeward voyage."[5] The complications here being financial: a fee for the pilot, a delay for the tide in getting the goods into port, a stay in port waiting for new cargo to be accumulated and loaded.

Unwilling to pay the pilot's fee or outwait the tide, the third solution was to hire lighters, a kind of barge pushed by tow boats to whose purpose was to unload and transport cargo from ships. Their flat bottoms would not be impeded by the shallowness of the passage through the sandbar.

The way through this Sandy Hook barrier was examined in 1802 by an early scientific group funded by the government, soon to be known as the Coast Survey. The naval officer in charge, Thomas Gedney, discovered a new way through the sandbar at high tide in 1808. (Not a foolproof choice, since it was only two feet deeper than previously, and ships had been known to sink in the channel as late as 1887.) The newly found depth that served well for many years, providing that the ship owners could find the channel outside of the entrance to lower New York Bay and would be willing to wait for the tide to lift their ships. (In 2001, a high-speed ferry ran aground in 4½ feet of water, 600 to 800 yards off Sandy Hook. The ferry had to wait until the tide freed the ship.)

The only way that the problem might be lessened was to take a few steps to bolster the safety of the port. Some of these steps included warning devices, the earliest of which seems to be the 1764 placement of a lighthouse placed off Sandy Hook, New Jersey, a warning beacon later followed in 1822 with the addition of a lightship. This location marked an entrance to New York Harbor, as did the 1828 construction of twin lights at Navesink, in the New Jersey Highlands.

## The East River Passage

So, if the Sandy Hook entrance presented problems, why not use the East River as the alternate route into and from the piers of the city? Taking this route, ships could steer out to the Atlantic many miles from the city and continue into Long Island Sound, for more than 100 miles in waters protected from ocean storms and rough water.

Perhaps steering a sailing ship between Martha's Vineyard and Nantucket Island, and having sailed the coast of Connecticut, at Throgs Neck, the vessel would enter the 13.7 miles of the East River. As part of the river's dangerous properties, little was known of its obstacles, even the ones that were visible, especially those underwater portions of the obstacles. (Rocks would continue to be discovered even forty years after the waterway was first surveyed.)

At the four-mile point on the river, the helmsman or pilot would maneuver around Riker's Island and soon enter a section called the Middle Ground, sailing in a southerly direction. For about just one mile before entering a section of obstacles, projections of land and even a whirlpool needed to be dealt with.

Here the vessel, having survived the Middle Ground, must steer a hard left to move into an area named Hell Gate where the ship must choose three distinct passageways to avoid nine rocks and two reefs. This is not to say the choices are easy because the decision of which passage is selected depends on weather, currents, wind, tide, and the skills of the sailors. The shortest route, named the Eastern channel, required that ship smoothly maneuver between Frying Pan Rock and Hallet's Point Reef. Having negotiated that space, the vessels must swing beyond Hallet's but not too far, because too wide a berth may send the ship crashing into one of the rocks of the Middle Reef, that reef fewer than 300 yards away to the east.

The helmsman or pilot must know as well that space between solid land and the navigable section of the channel is not the same thing. That is to say, that a channel marker, for example, will display where it is safe to sail, no matter the width of the solid land between. In this case, the land was Astoria and the rock named Bread and Cheese. And even once past that, the ship finds itself inside of a 600 foot space between Roosevelt Island (then named Blackwell's) and areas of Queens, a distance of about two miles. For those two miles, the 600 feet of space drops to less than 300 feet of channel.

Three hundred feet may sound like sufficient room, except for the famous turbulence of the East River. That area attracted the eye of Washington Irving, who wrote in 1824: "a very violent and impetuous current, it takes these impediments in mighty dudgeon; boiling in whirlpools; brawling and fretting in ripples; raging and roaring in rapids and breakers."

The other choices of routes were called the Middle Channel, which had more room to steer through Hell Gate, and the Main Ship Channel, the one closest to Manhattan which looked to have the most room to maneuver but at the same time would take the most time to navigate. And still have to steer a passage just as narrow as the East Channel as it moved south past Blackwell's Island.

Two hundred years would pass before serious and detailed surveys, currents and tides were begun on the problems facing shipping on the East River.

## Fulton on the Hudson

Even while the danger persisted on the East River, boat travel using the Hudson River persisted. After all, on the Hudson just west of Manhattan the space between banks measured 4,700 feet. The River's importance resided first as the waterway to carry freight and passengers to Troy and Albany. Reaching Albany, a route could be reached east to Boston by land and later by rail as well as partial and primitive routes west to Buffalo and the Great Lakes.

Years before, in 1807, Robert Fulton's *North River Steamboat* reached Albany, 150 miles north, in 32 hours with a passenger ticket costing $7. Passage on sloops to Albany took four days. Fulton, with his powerful partner (and uncle-in-law) Robert Livingston, followed that initial trip with a schedule of trips, the first reliable steamboat schedule. He and Livingston found that they were able to convince the legislature to grant them a

**Wikimedia provides a watercolor by Bard Bros. of an 1836 vessel. The painting shows a typical side–paddle wheeler of the time that was used for transport on calm waters, in this case the New York to Albany run.**

30-year steamboat monopoly for all of New York State. When the monopoly was judged illegal in 1824, many others rushed to take part in the business of servicing all the stops along the way and back to Albany. This does not mean that the Fulton and Livingston Line was out of business. In fact, by 1823, the Fulton and Livingston Line was running six days a week, alternating two boats every other day, one leaving New York City, and the other leaving Albany on a trip that cut the original time in half as more efficient engines were invented.

According to the New York State Education Department, "Within a few years, steamboats were able to make the New York–Albany run either overnight or by daylight, and this development resulted in two distinct styles of boats. Night boats offered larger sleeping cabins called staterooms. Day line boats, in contrast, were largely open. By the mid–1830s, steamboat traffic on the Hudson and across America's rivers, lakes, and bays had increased to over 700 boats."

Other destinations were established too. The man to be called "The Commodore," Cornelius Vanderbilt, began with a short ferry trip on the *Nautilus* to Staten Island in 1817. Soon he advertised trips from New York Harbor to Philadelphia. Ferries aided business with trips across the Hudson to Hoboken and regular schedules to New Haven followed soon after.

At the port of New York in 1820, goods worth $35 million left and entered. Ships carrying more than 248,000 tons used the port. Revenue for the entire country just from customs duties in New York reached $6,428,571, supplying a third of the federal income.

## Clinton's Ditch

Those revenue numbers were about to quickly be exceeded because of a New York State construction project begun in 1817, a project once labeled a "folly." Waterways named canals had been completed in France in 1694 and in Britain a national canal network was constructed in 1801. This new $7.6 million undertaking, the Erie Canal, extended 363 miles—the width of the state from Buffalo to Albany. Constructed with basic tools, a canal trip westward from Albany would take goods and people 62 hours to reach Buffalo.

Canals became popular. In 1828 the D&H Canal began to move coal from Honesdale, Pennsylvania, mines 108 miles to the Hudson River at Eddyville, New York. From there the coal traveled downriver to New York City. In 1829, Canada built the Welland canal enabling ships to take goods over Lake Ontario to Lake Erie and thus to Buffalo's harbor, a transfer point to the other Great Lakes, railroads, and the Erie Canal. The Erie Canal itself would be enlarged to 70 feet wide and 7 feet deep to accommodate larger boats of up to 240 tons.

When the canal was completed in 1825 with a legal speed posted at 4 miles an hour, horses pulled the canal barges along a towpath next to the canal.

As a bonus to all the other aspects of the canal, the path of the canal allowed travelers access to towns that stagecoach lines did not. But if the speed was limited, the sheer amount of grain, or lumber or anything else was massive. Simply, using the coefficient of friction, the towpath horse can pull more weight than a horse pulling on land because the canal's water greatly decreases the forces of gravity and friction. Not being pressed down into the road by gravity as a wagon would, the canal boat is floating and gliding as it moves.

Lake Erie and the rest of the Great Lakes and all the cities and ports on the Great Lakes were now connected. And there was more. Three other New York state canals were built in the same period: the Oswego, the Cayuga-Seneca and the Champlain and these three all connected to the Erie Canal. These three made it possible to connect directly with Lake Ontario (the Oswego Canal), connect with the Finger Lakes (the Cayuga–Seneca), and, shipping up from Albany connect with the Poultney River. That river flows into the very long Lake Champlain and at its northern end meets the River Richelieu, taking ships into the St. Lawrence River at Sorel. It's downriver to Montreal and easterly all the way to Newfoundland. A water route from Albany to Montreal was in that way established.

Once the Erie Canal opened, goods from Detroit, Cleveland, Buffalo and Syracuse (and Toronto) traveling to New York City (and from there the world) could all move by water. From all the ports on the Great Lakes, sloops and schooners could land their goods in Buffalo where they could be moved onto canal boats. Once in Albany, the goods again might be off-loaded to another boat or the barges could be pushed down the Hudson to the Port of New York. New York State had solved the problem of the lack of roads and in so doing connected itself with much of the northern United States. The canal sped up the time of moving passengers and material by 60 percent. The speed in turn cut travel over land rates by almost 90 percent, increasing the movable tonnage of merchandise at the same time. School children are taught that the cheapest route for wares to move from New York to Pittsburgh was to travel up the Hudson River, across New York by the Erie Canal to the Great Lakes, ending with a short overland trip to Pittsburgh.

Before the canal, shipping goods overland between Lake Erie and New York City cost $100 a ton and took more than three weeks. Afterward, the cost dropped to less than $10 a ton and the trip took just seven days. The canal was so popular that, by the late 1820s, up to $15 million worth of goods—such as flour, whiskey, and wheat—were transported each year. In addition, about 1,000 people a day traveled the canal. According to the Ric Burns' PBS documentary on the history of New York, a ton of wheat from Buffalo that used to take three weeks and cost $100 to transport, now reached Manhattan in seven days and only cost $6.[6] In 1834, 1,950 vessels entered New York harbor carrying 465,000 tons of cargo.

Even before (1829), just after the influence of the Erie Canal began to be measured, "the net revenue arising on foreign imports, for the port and harbor of New-York amounts to about $17,500,000, being full two-thirds of the whole revenue of the United States. A total of 1,634 vessels, from foreign ports alone, arrived at the harbor, made up largely of 757 brigs and 433 schooners, while the Erie Canal set afloat on it passageway 5,015 boats, carrying three million tons equaling $156 million each year—each day, 20,000 tons, 26 times the quantity carried by railroads. If traffic increased, shipbuilding changed and declined."[7]

But for those ships that wanted the most economical route to New England and even to Canada's Eastern provinces alone, the East River was the best route north. And from New York to Britain's major port, Liverpool, you must sail north. The East River flows north and then northeast.

## Shipbuilding, Commerce and New York City

On the land surrounding the docks of Manhattan and Brooklyn were found the businesses of the waterfront and it was said at the time the life blood of a nation is its commerce. Businessmen gathered at the Merchants Exchange building, rebuilt in splendor in 1835, at Wall, William, Hanover and Exchange Place. A kind of temple to commerce, the building would have cost more than $20 million in today's money. The goods were partly moved into the storerooms of the country through the Custom House, another similarly resplendent building of white marble and granite (eight years to build) at Wall and Nassau.

The port of New York continued to grow in wealth and importance. Shipowners whose vessels entered the port from the East River understood that they were able to send their goods up the Hudson as well as to the beginning of the Erie Canal too. Certainly, watches made in Connecticut, for example, could be conveyed all over the country much more rapidly than overland. Within 15 years of the Canal's opening, 1840, New York was the busiest port in America. New York registered more than 53 percent of all imports to the country. For businessmen of all kinds in New York City, more than 21,000 seamen came into and left the port of New York, all of them spending at least some money in the city. Clearly, the commerce of the city depended on the ships but also on the mariners. Undoubtedly, the port's incoming traffic was not limited to the cargo from the Erie Canal. The ships' capacities changed too. "In the 25 years after 1815 American ships changed in weight from 500 to 1,200 tons and in configuration from a hull with a length 4 times the beam to one with a ratio of 5½ to 1. The faster and thus shorter journeys meant that the shipowner could earn back his investment in two or three years."[8]

This striking increase brought on by the Erie Canal greatly increased commerce for all kinds of New York City businesses and increased opportunity for employment in the city, thereby adding to the population who would spend their wages in the city. They worked at the 3,387 manufacturing enterprises in the city.

## Famed Ships and Shipbuilders

Commerce in the city was known in the later 1800s for its manufacturing as well as the distribution of goods from the maritime trade. Established very early in New York were firms along the New York waterfront that tended to vessels that had to be overhauled and repaired. In 1824 the first American drydock was completed on the East River. On 10th Street,

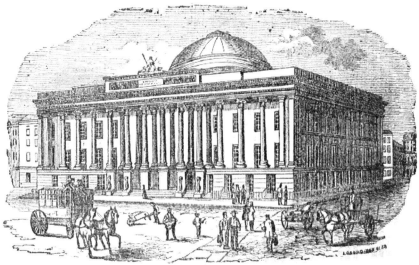

THE MERCHANTS' EXCHANGE.

**An 1835 image of the New York City Merchants Exchange Building, a kind of temple of commerce and wealth, much of the wealth due to New York City shipping. The building stands to this day at 55 Wall Street (courtesy New York Public Library Digital Collections).**

the New York Dry Dock took vessels out of the water for repairs, horses supplying the pulling power. Soon, an entire neighborhood would be called the "Dry Dock Neighborhood," from 3rd to 9th streets.

Many things were made in New York factories and small shops. For ship building and vessel repair and maintenance, the city had to produce sails, shaped wood, ropes, paints, caulk, and metal for the many necessaries on ships from anchors to pulleys.

As for the ships themselves, Morrison reports in his 1909 book, the output of the city's shipyards totaled in one year—1826—"twenty-three ships, three brigs, forty-nine schooners, sixty-eight sloops, twelve steamboats, fifteen towboats and nineteen canal boats, making a total of 29,137 tons."[9] Major shipyards were also established in New York, most notably the yards of William H. Webb and Jacob A. Westervelt. Webb's yard has been shown to have produced 135 vessels of all types in a thirty year period; Westervelt, 74 ships. "The following is the number, class, and tonnage of vessels built by Mr. Westervelt and his son, up to the present time [1860]: Fifty steamships, ninety-three ships, five barks, four brigs, fourteen schooners, one sloop, two floating light ships, one safety barge, eleven pilot boats, in all one hundred and eighty-one vessels, with an aggregate capacity amounting to one hundred and fifty thousand six hundred and twenty-four tons."[10]

**An 1870 photo of the East River waterfront.**

Located primarily on a space of a mile and a half along the East River shore of Manhattan, Webb, Westervelt, and other yards at one time employed over 10,000 workers: sailmakers, carpenters, caulkers, joiners, painters, coppersmiths, etc. Suppliers for those ships had to include the chart makers, and those firms that fashioned ships' instruments.

Soon the speed of these ships increased. "The Mayflower had taken 66 days to cross the Atlantic in 1620. By 1818 one of the first Sail/Steam powered packet ships—the Black Ball Lines—began to offer regular passenger service between Liverpool, England and New York. Soon that passage could be accomplished in 23 days from Liverpool to New York City. These ships weighed 300 to 500 tons. By 1845, Atlantic ships had doubled in size and were not credited as a success unless they had made at least a single east-bound dash of 14 days or less" wrote Robert Curley.

With the completion of the Erie Canal in 1825 the commercial greatness of New York was assured, and her Atlantic packet ships increased in size and numbers, averaging a thousand tons each in the zenith of their glory.

The year after the founding of Black Ball Lines, 1819, marked the first ocean crossing by a steam-propelled ship—a side paddle wheeler—in 29 days, 11 hours.

Always looking for a more economical voyage, Ships coming down the Hudson could save miles and time. Once the very narrow Spuyten Duyvil Creek and Harlem River be widened and dredged, a passageway to the East

River would be built by avoiding the distance and ship traffic in New York Harbor and sailing directly up into Long Island Sound. But surveys of these two waterways were few. The construction work on Spuyten Duyvil and the Harlem was delayed until 1895.

Even though by 1840 New York was the home of more tons of cargo than New Orleans, Boston, and Baltimore combined, the East River's problems were not even seriously studied until the 1840s. It may be that the lack of devices to accurately measure, for example, the depths of the rocks may have added to the problem, but not to be forgotten is the lack of money needed to complete the studies. Once the funds were forthcoming, the dangers of the Hell Gate looked even more serious than simple anecdotal evidence supplied by river pilots.

*Gothamist* sums up the situation this way: "By the 1850s, one in fifty ships passing through the Hell Gate were either damaged or sunk—an annual average of 1,000 ships ran aground in the strait.... Captains seeking to test their mettle would have to wait for the slim window of time where they could navigate their ships safely through the Hell Gate."[11]

Was overland journeying forgotten in this era of canal building? In most cases, roads might be laid out from city to city but no further. There were "turnpikes" being built. Figures show attempts at almost 1,600 of them but they had limitations. Besides continually being shown to be a bad investment, "the typical length [ran only] 15 to 40 miles."[12] Cheaper "plank roads" were built during the so-called "Plank Road Boom" of 1847–1853. However, problems remained. "Although plank roads were smooth and sturdy ... they lasted only four or five years.... Thus, the rush of construction ended suddenly by 1853, and by 1865 most companies had either switched to dirt and gravel surfaces or abandoned their road altogether."[13] The best roads—the best way to move goods and people—were still the lakes and rivers, plus the canals.

## Two

# The Business of the City

## *Some Studies Are Begun*

The East River began, finally, to receive some attention. The state's Chamber of Commerce declared "the immense value" of the shipbuilding industry in NYC remained "an active and permanent means of extending at home and abroad the commerce the manufacutures [sic] of this country can scarcely be overestimated." New Yorkers first pressured the U.S. government to do something about the East River waters in 1845 and repeated their requests often. They pleaded at least for the government to supply information about depth, tides, and currents. These requests for data, for measurements, often ended up in the hands of Prof. A.D. Bache, superintendent of the Coast Survey.

Bache was able to report on April 15, 1845, of surveys at Hell Gate completed by Lieutenant C.H. Davis of the U.S. Navy. In his report of February 15, 1845, Davis said "The dangers in this channel arise from the great strength of the currents, and the number and position of the rocks and reefs. The strength of the current is such that sailing vessels can only stem its force or escape from it by a commanding breeze." He also reminded the government of the military aspect of Hell Gate's passage: "The removal of the obstructions to the safe navigation of Hurl-Gate, is recommended by a regard to the future naval defenses of the country."[1] (For some years the use of the word "Hell" was determined to be wicked. Not so for Washington Irving: "Certain mealy-mouthed men, of squeamish consciences, who are loth to give the Devil his due, have softened the above characteristic name into Hurl-Gate.") As to the waters themselves the same article noted, "at Hell Gate the channel turns at right angles around Hallet's Point, Astoria, and the current runs with a velocity varying at different stages of the tide from 3 to 10 miles an hour over or around."[2]

Little was said about the effect of winds, due to their unpredictability. In a report to Eben Meriam for the Chamber of Commerce of the City of New York, Bache while on board the *Petrel*, recorded that 50 ships "had gone ashore" within his view in one month and even seven in just one day.[3]

Likewise, in 1850, another survey of the East River by another government employee, Lieutenant Craven, was completed and once again not only pointed out the problem in navigation but also suggested ways to lessen and remove the problems. At the same time as that survey was being read, the city could claim tonnage in the port of 931,193, representing 25 percent for all of the United States. In February, the New York State Legislature urged the federal lawmakers to make provision for the removal of the rocks "which obstruct and render dangerous ... the East River at Hurl Gate."[4] At least a lighthouse was erected far north of Hell Gate at Execution Rocks off Sands Point and funds of $309,837.49 were spent on docks and slips for new and repair work.

Yet much needed to be studied, surveyed, measured. How large were the rocks and reefs? What was the lowest point reached (called "low water") at various places in and near Hell Gate as well as in other places in the river? In response to urgent requests to state and federal governments from New York City maritime businessmen, other studies would be ordered in 1853 and 1856 but they were limited in scope.

True, the East River "is scoured by strong tides, which keep it permanently free from shoals of sand and mud, a scouring that makes the depth of the water conducive to forward progress."[5] But the rocks and reefs are not eroded, with Frying Pan, for instance, only at 11 feet, as measured at mean low water. In a collision with Frying Pan, the bottom or sides of the ship may be breached, split apart, and the costly vessel, the costly cargo, and even the crew might sink below the waters of the East River.

Even now the warnings exist: "'Currents often reach ... as much as 5 knots in Hell Gate. The tidal current in the East River floods North & East and ebbs South & West ... the combination of strong tidal currents ... heavy swirls and boils' will be experienced."[6]

## The Channels of the East River

These currents, swirls and boils affected every vessel that attempted to sail away from or into the piers and docks of New York City. Even though mariners could choose between three distinct "channels," the turbulence of the East River affected them all.

Bache and others had in hand some four surveys of the River. And though he had a clear notion of the position and rocks and reef up and down the river, the outcomes of a rapid current and water smashing into rocks could not be predicted. Water which flowed at 5 knots interacted with rocks so that eddies formed. What he did know was the powerful disruptive effects of the three significant waterways—water velocity, changing currents, eddies—combined with the tide from the Atlantic and the tide from Long Island Sound.

All of this can be best understood by following a ship coming out of Long Island Sound. For a long time, there was little difference in the capability of ships to deal with what awaited them and it did not matter if the crafts used sails, side paddle wheels, or propellers.

## From Throgs Neck to Little Hell Gate

At Throgs Neck, where the opening measures 3,400 feet, a ship traveling south to the New York City piers and docks enters the 13.7 miles of the East River (technically classified as a tidal strait) and finds more smooth sailing with wide passages on the way down to New York Harbor until it passes Riker's Island and makes its way, as it turns left, toward Ward's Island.

The ship must now navigate, first, through six rock obstacles named Way's Reef, Shelldrake, Pot Rock, Frying Pan, and then, closer to the shore at Ward's Island, Holme's Rock and Hog's Back. For the moment this is the sole navigation problem, except for the fact that a whirlpool, capable of capturing the vessel, forms around Pot Rock at various times of the day.

Opposite 125th Street in Manhattan, a minor body of water, Bronx Kills, extended eastward from the Harlem River to the East River, wide enough then at 500 feet, but of little use to most ships.

## Little Hell Gate

There the sailing ship entered a narrowing between Astoria, Queens and Ward's Island (now long since joined with Randall's Island), a space of about 1,200 feet. At Ward's, a ship met the first waterway it must deal with, the (now filled in) body of water then too big to be called a creek yet not big enough to be labeled a river, called Little Hell Gate. At 1,000 feet long, 165 feet wide, this narrow passage connected the Harlem River with the East River. The ship was now at about 112th Street, struggling against the forces of the Harlem and Little Hell Gate. (Randall's Island and Ward's Island would merge. conjoin, with the filling in of Little Hell Gate sometime in the 1960s and Sunken Meadow would be subsumed by Randall's.)

If not made helpless by the rapids and breakers created by the currents of Little Hell Gate and the Harlem River, the ship, depending on the

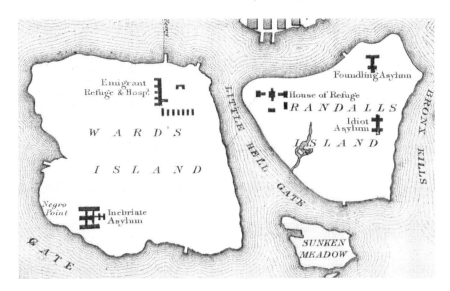

This 1883 map shows the placement of various hospitals and asylums on what were then two separate islands. Little Hell Gate, flowing between them, and Sunken Meadow, along with Ward's Island, by 1960 would all become part of Randall's Island.

time it chose to make its attempt at gaining the harbor, would also be at the mercy of powerful tides coming from Long Island Sound or the Atlantic Ocean. The tides that "meet and cross or overlap each other at Hell Gate" could raise the water at Hell Gate 5.1 to 6.6 feet.[7] So, water poured in, the tides raised the ship, and the ship's pilot found he is caught in conflicting currents that swirl at 8.8 miles per hour. The pilot might look south, ahead, and see a narrow piece of land, the 700-foot-wide Roosevelt Island, in the middle of the river, funneling the waters to the south even as the ship must enter one of the narrowest parts of the passage, should it safely avoid Hallet's Point Reef.

But no matter the width at any one spot, the ship must make its way somehow through a space of just 600 feet, 2,800 feet less than at Throgs Neck. The narrowness is partly explained by the fact that the river flows around Randall's and Ward's. Of course the space between bodies of land are not the same as a channel in the East River, "channel" meaning the deepest and safest passage in a broader body of water.

## Hallet's Point Reef

Approaching now on the mariner's left was a three acre point of land known as Hallet's Point Reef. This reef extended 720 feet along Hallet's point, like the blade of a partly folded pen knife, but what mattered to mariners was that it also extended 300 feet into the channel and, pointing north, acted as if it were a dagger stabbing at ships, or as an axe ready to carve them open, as seen at center of the map.

This projection of land caused two things. First, the helmsman or pilot knows he must avoid this pointed part of Astoria, while remaining aware that the evasion narrows the area that he can use to sail. Knowing the necessity for evasion, he employs his sails and his steering to move his ship out and away to the west.

The ship had now entered Hell Gate. If the ship's pilot was lucky enough to avoid powerful winds that day and perhaps, fog or rain or snow, he might find Hell Gate to be a true test of his seamanship. Though the ship may have survived past waters pouring into Hell Gate from Bronx Kill and Little Hell Gate, those two swirl in with forces from the larger Harlem River, whose width reaches almost 1,000 feet. The noise on board ship begins to pick up, just from the water alone. Should the reef at Hallet's be avoided by swinging wide of it, the current may carry the ship onto the nine acre Flood Rock, not far from Hallet's Point.

While the eastern shore of Manhattan may lie 2,000 feet away, the Middle Reef, those nine acres of sharp rocks, lay closest to Hallet's Reef being less than 700 feet of water away. So, what is needed here is to swing the sailing ship hard to the right but not so hard that it is crashes into the rocks. Once the vessel has avoided both Hallet's Point Reef as well as the Middle Reef, it can travel down the Eastern Channel on the east side of Roosevelt Island down to piers along the East River closer to The Battery.

Many other rocks lay in the ship's passage south in the river (23 or more obstructions lay in the East River) but those have always seemed to be less of a threat—once the money appears for them to be dealt with. If Hell Gate were improved, "the route to Europe would be shortened by fifty miles; the tedious waiting for high water at Sandy Hook would be avoided, and a full day's time gained on every voyage," so wrote the magazine *The Friend*.[8]

What was to be done to take advantage of these time savers and shortcuts? The question had to be answered simply because Sandy Hook seemed to hold little promise as a practical and efficient entrance into the harbor. The existence of a reliable, safe, entranceway was bound to affect the future of the city and country. But how could the East River be cleared? An enterprise of the kind so vast in scope and so difficult in character had never yet been undertaken. The problem was a new one to submarine engineering, with conditions of tide, current, and reefs so peculiarly difficult. If Washington Irving knew in 1824 of the disasters that awaited shipping at Hell

Gate, surely the city's merchants must know. When would the ship owners, the insurers, the merchant seamen take some action? How long before the ways to deal with the problem of the reefs and rocks might be found? How long might they wait before trying something to solve the problem?

Of course, ships continued to be damaged and sunk in Hell Gate, their cargoes with them. To give a comparison and using about 20 as the multiplier, in 1854 the ship *San Francisco* at the end of her maiden voyage was insured for $103,000 and her cargo for $365,000. When it wrecked, it was sold by underwriters for $12,000. The next year another vessel, valued and insured at $68,000, when wrecked was sold for $500.

The problems navigating Hell Gate were well known both in fact and fiction. In 1852, James Fenimore Cooper noted in his *The Water Witch* "the dangerous position of many rocks that are visible and more that are not, and the confusion produced by currents, counter-currents." He added the word "terror," to his description; the section was "*constantly* the cause of pecuniary losses" as well as "much personal danger."

## Changes in Ships and Ship Building

Too much money was at stake for nothing to be done. Money was collected, increasingly, by the federal government via tariffs and customs, money was being made and lost by insurers of ships, and merchants selling their wares to a wider and wider clientele. By 1850 this number of employees increased including coppersmiths, ship chandlers, ship joiners, shipwrights, edge tools craftsmen, and spar makers. The prevailing rate of wages had been very steadily maintained at $2.00 for a 10-hour day.

With the help of those workers, shipyards began designing and manufacturing more and better ships, faster, reliable ships. Located primarily on the East River shore of Manhattan, writes Harry Johnson, and Frederick S. Lightfoot, "these and other yards at one time employed over 10,000 workers. Local shippers could have ships of the highest quality built to their specifications, including 53 of the 116 clippers that were owned in New York."[9] Commonly constructed at the many locales were such vessels as packets, "a boat that carries letters, packages etc. and usually passengers at regular times." New York harbor was filled with a variety of other sailing craft. "There were smaller ships—brigs, barks and schooners—engaged in trade with overseas and domestic ports. They could bring sugar and molasses from the Indies; cotton, tobacco, lumber and marine stores from the South; or manufactures from New England—to name but a few of their cargoes. They might return with converted goods to the Indies and the South, and with cotton and imports to New England. On the Hudson River, the sloop was the principal carrier for freight."[10]

The first steamship built for regularly scheduled trans–Atlantic crossings is generally considered to be the British side-wheel paddle steamer SS *Great Western*, 1,360 tons, built in 1838. (The last of the paddlewheels would be launched in 1862.) Steamers though they may be, these ships were still wooden ships. Some may have been later plated with iron, such as the U.S. Civil War ironclads, but they still were equipped with masts. Not until the 1870s did ships begin to be built without masts, and that decision only affected a few boat builders. As late as 1899, the U.S. Navy was budgeting for ships with masts.

The steamships that still used masts and sails while using paddlewheels at the side or rear were making their mark. Though some saw the promise of advances in propeller design, the expense of steamers to build and to supply with coal might cause an owner to stay with wooden sailing ships. Ships in the categories of sloops, barks, brigs, and schooners—among many others—were listed in all large New York City newspapers, such as in 1840, when 5,614 ships arrived and departed using the port of New York.

The impact of ocean-going steamers in these years before the U.S. Civil War was blunted. To build an iron steamship would have required up to two years of work. Also, the long voyages around the Capes were too long for a steamer simply because the coal needed could not be contained in the hold. A wooden clipper could be built and at sea in less than six months. Charles Ingersoll wrote in 1852, "The Baltimore clipper, the American pilot-boat, the sea-steamer, and the pleasure-yacht, have all successively borne testimony that, in the construction and navigation of vessels, the builders of this country are not excelled."[11]

But the age of the clipper was foreshortened as more powerful steam engines and the propeller already were invented. With steamers came the need for better harbors to accommodate the greater size of these ships. The problems with New York harbor remained: the sandbar at Sandy Hook and the rocks and currents in the East River. After all, the displacement of Webb's ship *Celestial* reached 860 tons. But after the War Between the States, steamers (with masts), like the *City of Peking*, were found to be 250 feet longer than the clipper and weighing 5,080 tons to Webb's *Challenge's* 2,006 tons. Clearly, ships were growing in length and needing room to maneuver in New York as well as growing in depth of cargo holds, meaning the channels the ships sailed in from Sandy Hook or the Long Island Sound needed more depth than they had. The East River still had nine acre reefs and more than 30 other rocks that needed to be dealt with if the city was going to adjust to the newer, larger ships.

## The First Glimmer of Funds

On February 16, 1848, the federal government's Office of the Coast Survey wrote from Washington, "That it is very desirable to make such improvements as will lessen or altogether remove the hazard attending the ordinary navigation of Hurl Gate.... Of the sailing vessels that enter the Hurl Gate passage, it is estimated that one in fifty sustains more or less injury, by being forced by the violence of the currents on the rocks or shoals, and ... even steam-boats, with a motive power that keeps them under perfect control, and guided by the most experienced pilots, are not secure from peril."[12]

Something had to be done soon. The merchants worried about the tides and they worried about the reefs and they worried about the rocks and the channels for ships to move between the obstructions. In 1849, on July 20, a lighthouse keeper off Huntington, Long Island, counted 485 vessels sailing toward Hell Gate. Any one—or more—of those sailing ships might run into trouble: some might find themselves caught in the Hall Gate and have to be towed out. Some might go ashore and have to be somehow be moved from that position; at the same time, that stalled ship could block other ships from coming through the passages. Some ships might smash into rocks and be forced to jettison cargo in order to move the vessel. Some ships might even be ripped apart and sink, losing both cargo as well as the ship itself. That disaster too might block passage through the Hell Gate, delaying other ships. All of these instances cost money. A delay could inflate a crew's pay beyond what a merchant could afford to pay, for example.

In 1851 the Erie Canal now saw 5,015 boats on it, carrying three million tons, each day 20,000 tons. Each year the value of goods equaled 156 million dollars.[13] Meanwhile, nothing for the Hell Gate project came out of Congress' Committee on Roads and Canals (1831–69) nor the Committee on Commerce. However, in March 1851, despairing of any action by Congress, Henry Grinnell offered $5,000 for some sort of action on the East River's Hell Gate, and in the summer of 1851 proceedings were commenced for raising the required amount by subscriptions to a loan to be repaid by the U.S. Government.

## Coasters

The most famous of the luxurious steamers belonged to the Fall River Line, named after the city that became the terminus for the ships from 1847 to 1937. These luxurious ships, like others, ventured out into the ocean only on rare occasions, but instead rarely strayed far from the shoreline. In 1846

Richard Borden completed the Fall River Railroad, which enabled a connecting land route between Fall River, just south of Providence, Rhode Island, and other cities such as Taunton, New Bedford, Providence, and Boston. The Fall River Line was later purchased by the Old Colony Railroad, offering express train service from Boston to its wharf in Fall River where passengers boarded luxury ships to New York City. The train service used a locomotive, built in 1856.

*Harper's* described the passage: "toward five o'clock in the afternoon, is a sight which cannot be seen anywhere in the world. Between the low banks, and towering above them come the steamers which connect at New London, Stonington, Providence, and Fall River with the railway for Boston, their colors streaming, and the passengers promenading to the music of their string bands."[14]

The website mass.gov explains: "The most direct route to New York City from Boston involved a one-hour-and-forty-minute train ride to Fall River, followed by a nine-hour, overnight steamship ride to Manhattan" arriving at Pier 14 at the foot of Fulton and West Streets.[15] By 1860, maps showed wharves and piers from the Battery to 61st street on the Hudson River and from the Battery to 41st street on the East River. Other companies

**This 1851 Currier & Ives lithograph shows one of the beautiful vessels built by Webb and Hall. One of the larger vessels of its time, it accommodated 1,200 passengers on its route from New York City to Fall River, Massachusetts, flying the flag of the newly organized Old Colony Steamboat Company, a successor of the famed Fall River Line.**

would soon take advantage of transporting passengers from New York to Boston by combining steamers with railroads.

Of course, these steamers had to make their way up the East River to make the connections with Boston since the route meant sailing on Long Island Sound. But the East River remained perilous. The paddle wheelers were more expensive vessels and should any of them be wrecked in the maelstrom of the East River the expense rose even higher than the wreck of a sloop. Then too, there might be as many as 900 passengers on each of the steamships and possibility of the loss of life from that kind of wreck seemed to cry out for action.

Other steamers did not ignore the tourist industry and soon advertised for excursions were day lines and evening boats with sleeping cabins, with New York as the terminus. For example, in 1834, Daniel Drew formed "The People's Line" to operate steamers in direct competition to Vanderbilt and the North River Steamboat Association. By 1845 Drew gained control of the Association and the operation of most steamboats plying the Hudson River. A trip to Albany in safe and dependable ships meant people and goods could travel to that city, one of the largest urban areas in the country at the time. Railroads began to use Albany as a destination where passengers might find a route to, for example, Rutland in Vermont and the consolidation of what would become the New York Central had begun.

By 1849 there had been completed at least two studies of the situation at Hell Gate. But nothing had been done to address it. The Chamber of Commerce of New York City asked Congress for more money to clear Hell Gate. Part of that chamber would be men who owned ferries and proprietors of hotels at ferry landings, including a Mr. Dunlap, of the Hotel at Hell-Gate Ferry.

Ebenezer Meriman wrote to Benjamin S.H. Maillefert and in the month of October 1849, M. Maillefert, a French engineer with experience in Nassau in the Bahamas in blowing up a wrecked ship and some coral in 1847, came to the City of New-York (as it was then spelled) for the express purpose of making proposals to "remove the rocks from the channel of Hurl Gate by a new mode of Submarine Engineering. He proposed to break and scatter the rocks by exploding powder on the surfaces under water."[16]

There was no federal money at the moment for the work so Moses H. Grinnell, of Grinnell & Minturn, one of America's most important international shipping lines, and others said they were willing to contribute $6,000 to the single job of lowering the depth of Pot Rock to 24 feet.

The science of working underwater, as General Newton would later point out, was in a "crude state. The men then went down in a diving bell and drilled holes in the rock by hand and the explosion was made by [gun] powder."[17]

## *Maillefert at Work*

But with the money now being promised, as Newton later was to write, "a process of surface-blasting was first applied by M. Maillefert on August 19, 1851. His method consisted simply in placing upon the rock a charge of gunpowder, usually of 125 pounds, contained in a tin canister, and exploding it by means of the voltaic current."[18] Some of these targets for demolition could be seen above water and so the depth of water was claimed to be increased, on Pot Rock from 8 feet to 18.3 feet, on Frying Pan from 9 to 16 feet, on Way's Reef from 5 to 14 feet, and on Shelldrake, from 8 to 16 feet. Bald-Headed Billy and Hoyt's Rocks were blown into deep water. The depth on Diamond Reef was but slightly if at all affected, and no effect was produced on Hallet's Point Reef. M. Maillefert's operations resulted, using 620 charges containing 74,192 pounds of powder, but his gunpowder charges, at a cost of $13,861, had little effect below the water line.

Did Maillefert know what problems he was facing at, for one, Pot Rock? Had he read *The Merchants' Magazine and Commercial Review* in 1852, he would be alerted to

> The violent agitation of the water above and around Pot Rock, and the wild roar which accompanied it, was exactly such as if some sea monster were struggling in agony, vainly attempting to reach the surface of the water. When the tide was running, Pot Rock could not even be approached in a small boat, and the only available time for sounding the rock, or for blasting it, was during slack water, when the tide had ceased running one way, and until it commenced running in the opposite direction. But owing to the situation and character of the channel,

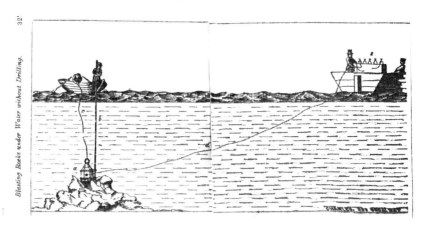

This drawing imagines how M. Maillefert's crude method of demolition of underwater rocks might proceed, although the work did not begin until August 19, 1851 (*The Farmer and Mechanic*, January 31, 1850).

slack water lasted only some few, never beyond ten minutes, and the operations were therefore confined to that limited space of time.[19]

Newton noted that Maillefert's operations that lasted from August 1851 to December 1853, supervised by Mr. Merriman and then by Major Fraser of the Army Corps of Engineers, with Maillefert doing the actual blasting by rowboat. The 243 changes on Pot Rock had lowered the rock 13 feet, with decreases as well in length and width of the rock. Maillefert claimed Merriman destroyed Bald Headed Billy and Hoyt's Rock, and lowered Pot Rock, Frying Pan, and Way's Reef. Though later these claims about Pot Rock and eight other obstacles later were found to be exaggerated, it seems likely that more than $13,000 had so far been paid to Maillefert. Under Major Fraser, Newton pointed out that though more explosions had been detonated, the gunpowder was being placed without actually looking at the site of the explosion.

It seems fair to characterize the work of Maillefert and Fraser as old technology. Their lack of meaningful and large progress in their work on the rocks of the East River stands in stark contrast to the inventions and discoveries of the second half of the 19th century, a period marked by the steam engine and the electric battery and advances in diving gear. Breakthroughs in drilling and tunneling were accompanied by the vulcanization of rubber and advances in hydraulics. The invention of nitroglycerine in 1846 had looked promising but it appeared to be far too unstable to be of any use.

## *The Politics of Funding*

A way of measuring the importance of smooth traffic in the East River is to understand that in 1854 there were 57 piers on that river, in a space just under two miles from the Battery. The last four piers on the Manhattan side were found at the foot of Jackson Street, south of Grand Street. (Some ferry slips were around the bend in the river.)

Nothing seemed to convince legislators—neither in Albany nor in Washington—even while New York in 1855 was averaging over one-half of the total foreign commerce of the United States to a total of $287,311,157. This meant revenue to the state and national governments arising on foreign imports of course. This figure does not include the commerce from inside the country of $261,382,960. The foreign commerce of New York comprises about two-fifths of the exports of all the United States, and somewhat more than three-fifths of the imports. Sometimes the ships did not make it through the roughest part of the East River.

An Advisory Council to the New York Harbor Commission, a council called the Board of Engineers, collected data in 1855 from 37 current and 10 tidal stations and then recommended eight demolition projects and four construction projects in the river. Its plan was substantially the same as one that naval Lieutenant Porter had proposed in 1848. In addition, "All of the plans so far proposed seem to have been limited to removing the smaller rocks lying in the channel; the possibility of clearing away the larger reefs was not yet conceived."[20] The council would have to wait over a decade for the work to begin. The New York State Harbor Commission wrote in 1856:

> The benefits of extended commerce, in augmenting the wealth and power of a people, in ameliorating their condition, diffusing the comforts of life, and enlarging the sphere of employment and usefulness, need not be portrayed. Provided by nature with a harbor unrivaled in geographical position, in immediate proximity to the ocean, and sheltered on every side from the influence of storms; of sufficient capacity to hold the navies of the world, and requiring none of the manifold contrivances of art for the protection of the ships that visit it.

To come to the point specifically about the East River, the council made its case for the benefits of the removal of the river's obstructions. The council members thought that such work, long overdue, would save lives and property caused by shipwrecks. The project too would facilitate both local commerce since East River ferries departed Manhattan at 23rd, 61st and 86th streets, as well as increase trade with New England and Canada.

Sandy Hook, full of steady amounts of sediment, did not lend itself to large vessels and the way via Hell Gate was shorter than through The Narrows. If the increasing profits continued through the increasing amount of tonnage coming to the port, the work must be done.

With the eyes of the city's men looking steadfastly on profit, the Board of Councilmen could speak in glorious terms of the business of the harbor. "The paramount interest of this City is its commerce. It is to that it owes its imperial proportions and its wealth. It behooves us, therefore, to guard … against all unnecessary burdens on the trade and commerce of the City."[21] The growing wealth of the city can be measured by the fact that in the 32 years from 1829 to 1861 the total number of shipping tons in the port increased from 1.2 billion to 5.5 billion.

The next year on the morning of September 27, 1862, in Hell Gate, the schooner *Peri* was sunk off Hallet's Point by a collision with the steamer *Island City*. Both ships, whether under sail or steam, could be to blame given the turbulence of Hell Gate. The navigation problems were ignored by the government. When a later request for funds came from New York City, the federal government requested more information.

# THREE

# General Newton Takes Command

As the 1860s passed, nothing of substance in work or funding lessened the problems in the East River. That the problems had been delayed and put aside because of the American Civil War was no revelation to ships' pilots, to ship owners, the insurance companies, and frustrated merchants. Thousands of dollars, thousands of pounds of gunpowder, and decades of petitions had accomplished very little, even as the harbor became ever busier over the decades. The *New York Times* reported on one day in 1865 "the following number of vessels in port yesterday, Steamers 49; ships 106; barks 131; schooners, 60."[1] Also in the harbor were ferries, rowboats, lighters, tugboats, fishing boats, boats for men is search of oyster beds. Even so, funding for East River projects had not yet become vital to state and federal administrations no matter the number of ships or the likelihood of wrecks and delays.

But pressure from various groups—ship owners, insurance companies, the Chamber of Commerce—began to be brought to bear. "It was natural that the masterminds of our commercial interests should covet the shorter and safer entrance through the Sound, so provokingly barred at Hell Gate."[2] After all it was believed that the clearing of "these obstructions once removed a hundred miles of exposure to a dangerous coast would be shunned" those hundred miles on the Atlantic route to New York. Also taken as truths were the ideas that "the route to Europe would be shortened by fifty miles; the tedious waiting for high water at Sandy Hook would be avoided, and a full day's time gained on every voyage."[3] In sum, continued the periodical *The Friend*, "the far-reaching importance of these advantages, and the possible effect of them on the future of the city and country are simply incalculable."[4]

It was also true that no one had come up with solutions to act on. The advancements and sophistication of technology for dealing with the difficulties of fixing the river's problems seemed not to have changed since Maillefert. In the fields of drilling and drill bits, in working underwater, and in explosives, changes were happening but still on an inefficient and

31

unreliable way. A West Point graduate, Alexander Dallas Bache, worked on surveying the harbor as well as overseeing the improvement of the New York waterways, usually for the Coast Survey. Hydrography might be able to measure and describe the physical features of the East River, but those data had nothing to do with demolition.

Part of the reason why so little had been finished lay in the technology of the time, two aspects of which were in submarine mining and blasting. Another part resided in the lack of a commitment from the state or federal government for funding the project. The third and last part needed to be the man with the right qualities of knowledge, patience, focus, and willingness to accept and use the new tools as they came available.

It was true that the federal government had spent money on the city. Between the end of the Revolutionary War and the end of the Civil War, defensive structures had drawn the most money. At least four forts at the entrances to the harbor and the East River were built under the supervision of General Joseph Totten.

Three more graduates would play an important part in the Hell Gate projects, all due to the foresight of the early superintendent at West Point, Colonel Sylvanus Thayer. In 1817 Thayer made civil engineering the foundation of the curriculum, with the first two years heavily stressing mathematics. Within a decade, the Army Corps of Engineers was responsible for building fortifications, roads and bridges improving the country's harbors and rivers. The United States Military Academy knew how important engineering skills would be to a military officer, a man very often the only person in a very large area of the country with skills in building.

Transportation and the life of the city were also in the hands of USMA graduates. From 1867 to 1871 George S. Greene served as the chief engineer commissioner of the Croton Aqueduct Department in New York. He also served as president of the American Society of Civil Engineers. Egbert L. Viele, another engineer-in-chief, headed up Central Park's construction in 1856, and Brooklyn's Prospect Park in 1860. Serving as Commissioner of the Department of City Works of Brooklyn, New York, in 1876 Henry Warner Slocum was appointed and was involved in advancing the cause for the building of the Brooklyn Bridge, and for streetcar systems in Brooklyn.

Throughout the 19th century, the graduates supervised the construction of fortifications all over the young country and were responsible for many maps as well as for road building and bridges. It became part of the mission of the Corps of Engineers to construct lighthouses, help develop jetties and piers for harbors, and carefully map the navigation channels. If a particular student showed promise in engineering skills, he might be retained at the United States Military Academy at West Point, New York, after graduation, to serve as an instructor.

## *Newton Is Chosen*

One such able student was John Newton of Virginia, son of a congress-man, the man chosen to be in charge of the Hell Gate obstacles, and much, much more. A graduate of the USMA in 1842, Newton was appointed the next year to the school as an assistant professor of engineering at age 21, leaving that post in July 1, 1846. Soon Newton's background included supervising engineering work on forts and coast defenses, dredging rivers from Maine to Florida, surveying in Mississippi and Texas and planning lighthouses in Michigan. He served as one of the commissioners for the improvement of the harbor of Montreal.

More importantly, he was familiar with Manhattan's harbor since ear-lier in his career, in 1860, he was appointed to be a member of the Special Board of Engineers for modifying plans of a projected Fort at Sandy Hook, New Jersey, and for selecting sites for additional artillery batteries at Fort Hamilton, New York.

In that same 1860, 3,982 vessels entered the harbor carrying 1,983,000 tons of cargo, so the average tonnage per ship entering the harbor increased from 238 to 498 tons of cargo in a 26-year period. More tonnage being car-ried also meant that the ships moved lower in the water and needed more depth in the New York harbor channels it used to navigate. Knowing the tonnage continued to be an important metric. On June 8, 1860, a powerful member of the city's chamber of commerce,

> Mr. Samuel B. Ruggles, reported that the number of vessels passing Hell Gate during the month of May last was 5,523, which, at an average of 150 tons each—a low estimate—would show a monthly movement exceeding 820,000 tons.... At the rate proposed by the Hell-Gate bill [H.R. 287 by Mr. Haskin] of one cent per ton, this would have yielded $8,200 per month, or $98,400 yearly.[5]

During the Civil War, the city paid $68,744 merely for the partial removal of just one obstacle—Coenties Reef.

Like others in the Corps, Newton disseminated his own engineering projects through a mandatory yearly account, beginning in 1807, called the *Annual Report of the Chief of Engineers to the Secretary of War.* (This report appeared under different titles in many different government publications such as the *United States Congressional Serial Set.*) In these annual publi-cations, work on projects and results of the Corps were described in min-ute detail. Made available to congressmen and members of the Corps alike, the annual reports contributed to both military construction and works "of a civil nature." They contained a description of projects like lighthouses, dredging harbors, and other maritime work and their budgets were down to the penny. They not only gave amounts of tools used, supplies needed, and equipment necessary, but more significantly they stood as descriptions of important scientific experiments, the testing of new chemical compounds as well as new methods of approaching tasks on the water. Taken as a whole all of these reports constituted a kind of magazine of breakthroughs in engineering.

Much of Newton's assignment would focus on the part of the river opposite streets in the 90s but his supervision and advice would be required for hundreds of miles of the Hudson, Lake Champlain in Vermont as well as any important engineering project, such as the Brooklyn Bridge, first called the East River Bridge.

Regarding his service in the Civil War, after Antietam, Newton was promoted to major general of volunteers on March 30, 1863. Stephen Sears described him this way: "General John Newton looked the very picture of a professional soldier. He was forty, tall and stiffly erect and with a determined look about him."[6] Mustered out as a "brevet" general of the Army of Volunteers on January 31, 1866, he was commissioned Lieutenant-Colonel of Engineers in the regular service at a salary of $4,000 in 1865. But the title of General stuck as far as the press was concerned.

The pressure from various New York City sources finally produced some results through the office of Brigadier General Andrew A. Humphreys, who was then the Engineer in Chief of the U.S. Army Chief of Engineers. In April 1866, Newton's orders arrived though he was on leave until August 20, 1866. Of all of the staff in the Corps of Engineers, John Newton was selected to be in charge of work centering in New York City, the most powerful city in the country, the richest city in the country, the city whose commerce through customs revenue maintained a large contribution flowing to the federal government. Once in this office in New York City, "he made an examination for the improvement of the navigation of Hudson River, the appropriation then being sufficient only for the repair of the dikes already constructed; but his report covers the whole ground, and the scheme then proposed is that which has been carried on, latterly entirely under his charge."[7]

As for his assignment, it is tersely stated as "Removal of obstructions in East River and at Hell Gate, N. Y., July 31, 1868, to Dec. 31, 1885." What Newton was about to take on was the entire passage of the East River for the commerce of both the Port of New York and of the entire country itself. This can first be shown by New York's contribution to the United States in three ways.

In 1866, $132,000,000 was collected in customs duties while the U.S. government owed $133,000,000.

Second, in 1866, the strength of the port's activities was illustrated by the fact that the country's imports and exports reached just over one billion dollars.

Third, in the matter of customs revenue, in 1866, the entire country (North and South; the war over) generated $179 million. New York, by itself, took in $132 million. Thus, 74 percent of customs revenue came from the port of New York.

Newton knew very well that his job would begin by being focused on obstacles in and around Hell Gate, but he always knew that so much of the financial health of the country depended on the execution of his duties as the chief engineer.

In the same month that Newton was ordered to report, a warning appeared from the *New York Herald* about Alfred Nobel's new experiments with nitroglycerine, a compound that promised spectacular results, while remaining a fearful substance in the eyes of many. "It is the duty of Congress to make the introduction of this murderous compound, without due safeguards, a crime punished with the utmost severity."[8] On April 24, 1866, the *New York Times* reported that the House of Representatives set to work on a bill "providing against the importation, transportation, and manufacture of nitroglycerine in the United States." The impetus for this legislation may have been the catastrophe at Aspinwall in Panama. *The New York Times* reported "Twenty-six Persons Known to be Killed and Twenty Missing" in addition to the destruction of the iron steamer carrying the explosive. The disaster led the *Times* to label the cargo "nitroglycerine, the infernal compound."[9]

It appears Newton was interested in nitroglycerine from his reading in technical and popular magazines. Taking command of a large part of the city's maritime life and its defense, even before the end of 1866 Newton was ordered to look into a demolition job before the year ended. He wrote: "December 18, 1866. General [Andrew A. Humphreys]: I have the honor to forward herewith a chart showing the position of the vessel wrecked off Sandy Hook, northeast from the light, and referred to in my letter of October 24 last.... The ... removal ... may be done with an expenditure of from $6,000 to $8,000."[10] Newton would take on that job the following August.

General Newton had moved into his office in lower New York at No. 7 Bowling Green, New York, headquarters in the city of the United States Army Engineers. Quickly the engineer reported his estimate for work on the Hudson between Troy and New Baltimore for dikes, dredging, revetments, etc., at a cost of $862,297.75. Then, he set to work on a detailed plan for the destruction of the obstacles in the East River, most especially in and around Hell Gate.

Forty-two years after Irving first wrote about the East River, many city politicians seemed finally to have been heard. Newton is first mentioned in an article from the *New York Tribune* on February 2, 1867, in which the general commented on the advance of various items diminishing the size of the harbor.

## The City Changes

Even as the General began his work on the estimate of his labors, the city was changing in all sorts of ways. For example, in February 1867 the two-centuries-old pear tree that Peter Stuyvesant planted on the corner of Third Avenue and 13th Street died from time and weather. Also in 1867, John Roebling drew his plans for what was called at the time the East River Bridge. Like Newton, Roebling's son and chief assistant served as an engineer during the Civil War. In a way, the Roeblings would soon be in the construction business on the East River while Newton would be in the destruction business in the same place. Two years later, Newton was appointed with two other Army engineers to examine the design of this new Brooklyn Bridge (independently of the civilian board), particularly as it might impede navigation.

Seeing the city migrating north from the waterfront, Delmonico's, on Chambers and Broadway, the city's best-known restaurant since 1856, in 1862 opened a branch in a three-story building overlooking Union Square on 14th Street and Fifth Ave, two miles north. There was centered New York City's first commercial theater district. Forgotten now, but then a source of great pride to New Yorkers were the harbor's oysters. Bedloe's (Liberty) Island and Ellis Island were called Greater and Smaller Oyster Island for many years.

The May 25, 1869, *New York Herald*'s vision of Manhattan remained limited, writing, "There is a point beyond which Broadway can never become a very great thoroughfare and that point is Thirty-Fourth Street...." A newspaper commented that "the social center of gravity has been changed in forty-five years from Wall Street to Madison Square."[11] And when that happened, hundreds of thousands of bodies from the Madison Square Park on 23rd Street and Reservoir Square (now Bryant Park on 40th Street) were relocated to potter's fields to Ward's Island.

To get away from city life and for picnics, a boat would take people to 59th Street. As the city grew (86 percent 1840 to 1850; 68 percent 1850 to 1860), the city's importance to the nation did not change. New Yorkers best understood it. William L. Stone's (1872) *History of New York City* asked, "Look at our city now in its extent, population, wealth, institutions, and connections, and consider how far it is doing its great work, under God's providence, as the most conspicuous representative of the liberty of the nineteenth century in its hopes and fears."[12]

In 1867 a sign of the city's "great work" could be found in the first elevated transportation line constructed by the West Side and Yonkers Patent Railway Company (using a steam locomotive) along Greenwich Street and Ninth Avenue. The local paper thought

> Another curious mode of transit is the air line [the current words for "elevated"] of railway. Some twelve feet above the ground, supported by a single row of pillars, and carried on cantilevers, a single line of a very narrow-gauge railway runs above the heads of the pedestrians on the pavement, and just on a level with the first floor windows. It is singular in its appearance and certainly does not add to the beauty of the thoroughfare.[13]

**A portrait of General Newton probably taken around the Civil War's Rappahannock Campaign fall of 1862 (Library of Congress).**

## Ferries, Pilots, and The News

As a clear example of the changes taking place in technology must be a movement away from horses, a transformation in the city transportation that has for the most part been long forgotten. It has to do with ferries. Writing in *The Chronicle of the Horse*, Sid Perkins explains that horse-powered ferries, or team boats once carried goods and passengers across New York harbor, in the 1814–1824 time period. The horse, on a treadmill, would be the motive power for a paddlewheel on a ferry able to handle 200 passengers in the late eighteenth and early nineteenth century. In fact, there are records of regular horse ferry service until 1824.[14]

However, steam ferries, though much more expensive than sail ferries to build and to fuel, soon became the choice of harbor bound New

Yorkers. The clearest picture of this new technology does not ignore the use of sail to serve as a delivery service. But given the speed of the current and the strength of the tides in the East River, steam often had to be the answer. Soon appearing was the man who would influence the city in powerful ways, Cornelius Vanderbilt. Starting with a $100 investment from his father, the soon to be named "Commodore" bought a sail ferry for service to Staten Island. With the increased business of the forts around New York during the 1812 war, Vanderbilt bought more schooners and expanded his business into New England and as far south as Charleston, South Carolina, as well, with a partner as he switched to steamboats, running a regular schedule to Philadelphia.

Then, on his own, Vanderbilt began service on the New York City to Peekskill Hudson River route. Peekskill had a railroad terminus for Albany, and so Vanderbilt could move freight and passengers a third of the way to Albany. (No bridge would be built further south on the Hudson than Albany until about 1890.) By 1869, trains would carry freight directly into the city.

This does not mean a decrease in ferry traffic in the city. In fact, by the time the Civil War started about 70 ferryboats steamed around the harbor. No bridge yet was built. But there were still large problems to deal with. Nature provided one problem: early in 1866, the East River froze solid and the Hudson was also blocked. "The brave-hearted," wrote the *Times* in 2003, "crossed the frozen river on foot to get to work, but most people stayed home. Stores ran out of goods, and businesses stood on the verge of collapse. It was jokingly said you could get to New York faster from Albany than from Brooklyn."[15] This would not be the last time General Newton would confront a frozen New York.

Problems arose too concerning pilots, those mariners who were paid to take command of a vessel to see it safely through into and out of New York Harbor. How much they ought to be paid, what tonnage of ship needed a pilot, and how far up the East River their province extended would be argued about for many years. Some ships ignored the pilots but at least one owner was arrested for taking a vessel through Hell Gate without a pilot.

Pilots knew that the East River had a major virtue that Sandy Hook did not. Because of the rapidity of the water flow in the river, channels existed to accommodate the minimum size required: the width of the channel needs to be at least four times as wide as the vessel using it. As for Sandy Hook,

> The approaches to this entrance are stormy and perilous. The entrance is obstructed by a sand-bar, over which vessels of large draught cannot cross except at high tide, causing constant, vexatious, and expensive delays. The inner

channel is crooked, shallow, and subject to shifting shoals, which make the passage uncertain and troublesome, if not dangerous.[16]

To aid shipping, New York newspapers such as the Bennett family's *Herald* and Horace Greeley's influential *Tribune* began listing the time of highest flood tide each day at Hell Gate. Newspapers too began to pay more attention to the traffic in the harbor totaling up, as they chose to classify them, each Brig, Schooner, Steamship, Bark, and Ship. Paying attention to each craft was one more way of demonstrating the commercial importance of the maritime trade in the harbor and rivers of New York City.

How important? During the financial year 1858–59, there were cleared (exited) from the port of New York 3,086 vessels with a weight of 1,476,279 tons, while the total that entered the port, foreign and domestic, 3,902 vessels of 1,890,144 tons, More than 700 revenue clerks processed these ships, demonstrating that one more group of workers dependent on the ease of traffic in the port. By 1860, one-third of all ships from foreign countries and 60 percent of the tonnage entering the United States entered through the port of New York.

But the port still had its East River problems for Newton. One newspaper, *The New York Times* on August 19 stated clearly, "It is … a matter of public notoriety, that dangerous reefs like those at Hell Gate exist to-day in the harbor of the chief commercial City of the Union. Every day or two we hear that some vessel is stranded on the rocks that underlie the East River entrance to the port." For example, vessels that "swung wide of Hallett's Reef to avoiding crashing by impaled on it were in danger of being driven by currents on to … Flood Rock." If New York harbor were ever going to be a smoothly functioning port, then the outbound passage through Long Island Sound and then Hell Gate in the East River must be undertaken.

Proof of the East River's dangers each year abounded. On August 18, 1866, as Newton was writing in his office, the steamer *Newport* of the Fall River Line, a vessel measuring 361 feet long by 43 feet wide, encountered the tugboat *Quickstep* pulling the coal barge *Jay Bird*. Unable to safely pass by each other, the tug and the barge's resulting collision sank both. The dollars lost from the sunken tug, the sunken barge, and the drowned coal barge plus the price of the coal were not reported, but no crewmen were drowned.

## Newton Designs the Projects

The General continued to read reports (one called "hydrographic surveys") on Hell Gate conditions that were decades old, reports such as the various United States Coast Surveys. This government survey group, founded in 1807, produced nautical charts, and, among other achievements, designed tidal predication machines found in Henry Mitchell's 1869 "Report on the Tides and Currents of Hell Gate." When Newton lowers and installs "fenders," and destroys the rocks and reefs, the impact of those tides will be diminished. Some of the surveys were vague, naming the river's obstacles only as "[r]ocks designated in earlier reports" or to be "removed as the necessities of navigation demanded."

The more troublesome of the rocks described in the study included these rocks identified as "the rock off Negro Point, the rocks near Woolsey's Bath House." More familiar obstacles included Bald-Headed Billy, Pot Rock, Shelldrake, Frying Pan, Holmes' Rock, Hog's Back, Heel Tap, Greater and Smaller Mill Rocks, Scaly Rock, Charlotte Rock.

Noticeable on maps were two large problems though they seemed to have been ignored, possibly because they seemed so impossible to conquer. The smaller of the two identified was called Hallet's Point Reef, totaling three acres. And even larger,

> The great reefs in the middle of pass—Flood Rock, Negro Head, the Gridiron, and Great and Little Mill Rocks—have been little noticed in any scheme of improvement, notwithstanding they form the greatest obstacle to the pass of Hell Gate being converted into a highway for commerce. In the existing state of the art of removing rock under water, the imagination was appalled at the notion even of meddling with such means.[17]

This discovery of how the Middle Reef was composed would lead to a project that lasted nine years. Obstacles called Reefs included Rhinelander's Reef, Diamond Reef, Coenties Reef, Ways Reef.

Newton was aware that if the rocks were lowered, or just their configuration changed, those alterations alone would change the flow of water in the East River. A change in the number of eddies, for example, might result and whirlpools might disappear or at least lessen in power.

By the time the General arrived in 1866 there were, at minimum, more than 20 learned and popular journals of science. Ordinary magazines, fascinated with inventions, began taking note of the latest patents, while more technical magazines flourished both for the lay reader—*Popular Science* and *Scientific American*—and for the scientific community both in Britain (*Minutes of Proceedings of the Institution of Civil Engineers* 1837) and the United States (*Engineering and Mining Journal* beginning in 1866).

All aspects of practical and pure science had their own publications such as the fields of chemistry, engineering, and technology. Learned societies had been formed in America, among the government, civilians and the military. Newton also had the assistance of G.W. Blunt, commissioner of pilotage, for reliable information. True, this was the era of invention, of time saving and labor-saving contrivances, as well as a much finer understanding of the forces of nature. For example, electromagnetism was more thoroughly thought out by Michael Faraday in 1831, and Morse had mastered the telegraph five years later.

Nevertheless, it must not be forgotten that the tools with which Newton began his East River work in 1866 were rudimentary. Gunpowder, hand drills, fuses (the type of safety fuse invented and patented by William Bickford in 1831), rowboats employed as drilling platforms, and underwater work from diving bells. A sizeable step forward was taken in 1859 with the invention of the lead acid storage battery "wet cell," a rechargeable energy source. The intricacies of electricity had just been discovered. But as each new advance became known, Newton did not hesitate to immediately employ them. He was a true scientist, trusting in what had been proven, even if still frightening to the general public.

Maybe some new metrics and some new technology, some inventions, might yet help with the problems at Hell Gate. Sometimes simple necessity did mother invention. When rail miles doubled the number of canal miles, tunnels were needed to complete trackage. In 1833, the first rail tunnel, named the Staple Bend tunnel was built in Pennsylvania. This 901 foot project took 17 months to finish at an average rate of 53 feet per month. Virginia's so-called Crozet tunnel with only hand drills and black powder dug though solid granite at a rate of 610 feet a year. Finished in 1858, it would extend 4,273 feet. One tunnel, the Hoosac, would play a part in the kind of demolition needed for the East River.

Laboring under the charge from Washington, Newton understood that he had the authority to begin the complex and puzzling Hell Gate work "in obedience to the provisions of the act approved June 23, 1866, making appropriations for the repair, preservation, and completion of certain public works." It read in part, "the Secretary of War is hereby directed to cause examination or surveys, or both as aforesaid, to be made … at Hell Gate, New York … [and] For improvement of navigation of the Hudson river, fifty thousand dollars."[18] There were many other New York and Vermont projects specified in the act, but those others did not draw much attention to begin with.

Nevertheless, the general's projects would extend, in the north, from Goose Bay, the northernmost harbor on Lake Champlain, to 300 miles south at Staten Island, and from the Battery in southernmost Manhattan to the end of the Hudson River at Henderson Lake a space of 270 miles, taking his supervision to within 66 miles of the Canadian border. It was water that carried the commercial interests of the nation and it was water that needed to be tended to. The changes to the Harlem River and to Spuyten Duyvil (where the Harlem meets the Hudson) would also be begun by the general and completed in his last year.

Most New Yorkers in the maritime trade knew that in Hell Gate's area, "More harm is suffered and more risks incurred here in a space of 2,000 yards, than in all the rest of the navigable waters this side of New York to the farthest extremity of the Sound." How much cost in time and money lay in those 2,000 yards of Hell Gate? The writer continues "in the channel, a thousand vessels a year were wrecked or seriously damaged by collision with its projecting rocks … it is no uncommon thing to see two or three vessels go ashore on Frying-Pan, Gridiron, or some other of its treacherous reefs, in the course of a single day."[19] Newton would spend almost two decades on those Hell Gate dangers.

In 1866, more than 15 years had passed since Merriam and Maillefert first attempted to lower the rocks and reefs. Maillefert hypothesized that by placing an explosive on top of something—a wrecked boat, a rock—the blast's power would move downward and thereby demolish his objective. This speculation seemed to be proven for Maillefert in Nassau in the Bahamas when he blew up a wrecked ship using this theory. But hard rocks were not wood ships. At least, Maillefert's work in the early 1850s gave Newton some glimpse into how difficult his assignment might be. Newton, then 44, wrote in a report dated January 21, 1867, "Any reasonable hope of effecting this improvement depends upon the rejection of misty and fanciful schemes, which cannot be brought within the rules of ordinary calculation."[20] Fanciful scheme or not, certainly Maillefert's method proved impractical. Newton's plans were anything but "misty and fanciful" and depended on "calculation," experimentation, and a reliable staff.

Newton's engineering instruction, experience, and scientific reading had taught him five clear things fairly quickly. First, the job in front of him was complicated by competing problems. Maillefert, it turned out, had many vocal supporters in the city who wrote to the area newspapers in support of the Frenchman and his methods. After all, they could see the explosions that Maillefert set off, no matter how little those explosions accomplished toward reaching the desired 26-foot depth level.

Second, the blasting begun 16 years earlier had changed very little in lowering a few rocks and thereby the depth of passageways—channels—for vessels. Newton observed that Major Fraser had stated that a few rocks had been blown into deep water and a few lowered a small amount—for $20,000. It appears it was common knowledge that Maillefert's company—three men—had pocketed almost the salary of the President of the United States.

But something changed by 1854 and for some time Maillefert's participation of the East River disappeared. The government, rather than paying him, appropriated $64,000 for more surveys. (But Newton knew the truth and wrote it: "The improvements at Hell Gate were made by exploding charges of powder placed upon the rock, no advantage being taken of submarine apparatus of any kind to establish these charges where the greatest effects might result."[21])

Thirdly, he knew that "The precise depth to be attained will be dependent not only upon the present, but upon the future, draught of vessels."[22] The combination of the growing abilities of ships' architects combined with increased traffic of steamboats had to be considered. Steamships, particularly powerful ocean-going steamships, began to appear at the end of the 1860s. They weighed twice as much as the largest sailing ship and carried in their holds many times more cargo.

Fourth, explosions could not be detonated from rowboats, as Maillefert had used. There was little stability in the currents and tides of the East River with something so small and light.

Fifth and lastly, the General knew that new methods, new tools, were going to be needed if this job was to be completed. One of those methods would have to be some way to efficiently work underwater. The diving bell's nature limited it to shallow depths, since the depth determined how much air could be contained inside. Any improvement in the diving bell would be limited by the vessel from which the bell was lowered. Newton deemed the bell "impracticable because of the certain disturbance of the apparatus by currents, and the liability to collisions from passing vessels. All of the plans so far tried or proposed seemed to have been limited to removing the smaller rocks lying in the channel."[23] As for the two larger reefs, at the minimum, advances were needed in explosives, in the tools for mining, in diving equipment, and thus in the ability to work underwater, at the root of the obstacles.

One way of understanding the danger that lay in the influence of, at minimum, 30 rocks and reefs. One of the larger obstacles, but not the largest, was named Frying Pan Rock and was located halfway between Hallet's Point and Ward's Island. At 12,475 square feet, the rock weighed 139,522,500 pounds, and swirling around that rock was the speed of waters affected by tides, by currents, by wind. Some water would flow past the rock from the Harlem River, some from Little Hell Gate, some from Long Island Sound, and some from the Atlantic. Mixing and mingling, circulating and swirling, the water moving around the rock would cause even more turmoil in the East River because of the phenomenon known as eddies. While it is true that a rock might seem to slow the movement of water—its flow velocity—at the same time, the rock is disrupting the flow so that the water behind the rock runs counter to a main current. This counter movement therefore becomes circular, a swirling vortex that pulls whatever is caught in the whirlpool downward.

There are often mentions of whirlpools to be found in the official reports in the larger Hell Gate area, almost as if the presence of whirlpools is ordinary, particularly the one near Pot Cove. The problem of these vortices extended, of course, to vessels making their way up and down the river, but created difficulties in the task facing those workers employed by Newton in moving alongside of the rocks and holding steady in the current and its varieties.

The hard, pointed and sometimes hidden rocks in the East River endangered the regular arrival of coal, lumber, food stuffs, etc., as well as the safety of passengers, particularly to and from New England. A few of the many rocks and reefs sometimes were clearly seen, as at low tide, but mostly they lay below the water line.

## Some Solutions, Some Problems

Newton would have to find a reliable way to combine electricity to set off the explosions with some yet unknown material more effective than the five tons of gunpowder Maillefert employed, and a more reliable fuse as well.

He would have to contend with the eight mile per hour currents, the tides, the whirlpools. He would have to contend with the hardness of the rock, since much of the material to be demolished turned out to be gneiss, hornblende gneiss, a layered metamorphic rock, possessing a hardness of 7 due to its high quartz content. On the Mohs Hardness Test, only topaz, corundum, and diamond were higher.

For the many smaller rocks and reefs in the river, Newton considered essential that the drill and its bits be firmly controlled by being tightly confined to avoid the swirling of the currents. Perhaps some iron tubes originating from a stable platform, certainly not a rowboat. He insisted "that the divers or the machinery necessary to handle and remove from the bottom the rock blasted should be protected from violent currents."[24] How the drilling and the divers might be accomplished remained for a moment outside of current means.

To equip his workers and their jobs, Newton knew he needed current breakers for the divers, floating derricks to lift the debris, 24 scows of varied sizes to receive the debris, and anchors and chains to steady the drilling platform and money for losses and maintenance.

To see his work as comprehensive, and so taking into account previous schemes, surveys, and measurements, the general looked at the many obstacles all around Hell Gate as well as in Hell Gate itself. Sea walls and beacons for other points would aid navigation and lessen the danger for vessels.

At this point in judging the size of five rocks—Pot Rock, Frying Pan, Way's Reef, Shelldrake, Scaly Rock and Heel Tap—Newton would be tasked with demolishing and then having the remnants, 6,680 cubic yards of rock totaling 1,706,994 pounds, grappled and taken away in scows, boats under contract to the government. Where the rock once dredged might be deposited also must be determined.

But just as important as the mid-river rocks, there remained the jagged extension of Hallet's Point Reef, and the nine acres of what would was called Flood Rock. The destruction of Hallet's, the reef that projected out into the river, had been insisted on as far back as 1848 by Professor Bache. The two major obstacles, that created not only dangers to shipping, but of themselves creating the eddies that might push or drag ships into danger. Those two reefs, with the information Newton possessed in 1867, would require the demolition and hauling away of 137,700,000 pounds once Hallet's was destroyed and 947,562,300 pounds of rock once Flood Rock was eliminated.

Dismayed Newton may have been, while understanding the importance of doing what was asked of him and completing the tasks in the safest and most cost-efficient ways. Therefore, Newton's report of January 1867 used his engineering experience in locales all over the country and Canada. This yearly report for fiscal year 1867–1868 estimated the dollars for the labor of carpenters, blacksmiths, divers and more. The cost of each gunpowder charge included the cost of surveying the rock, drilling the rock, of the explosives canister and charge, the pay for divers in placing the charge, the services of the operator and boat's crew in firing, the expense of wires.

As an experienced engineer, he understood that the small explosive charges are first placed in the natural or drilled holes in the gneiss to split up the rock; the larger charges are seated into the crevices to finish the work of breaking up the mass. He estimated the cost for the explosives—and here can be seen something new, estimates to the penny: for one 50-pound charge of gunpowder of $45.80, drilling included.

## Newton Includes Nitroglycerine

And then followed an "estimate for 5-pound charge nitroglycerine $43.04, drilling included; 'estimate for 15-pound charge nitroglycerine $35.35.'"[25] Around 1860 Alfred Nobel carried on experimenting with nitroglycerine in St. Petersburg. He finally got nitroglycerine to explode underwater, by surrounding a glass tube full of the liquid with gunpowder in a zinc can, and using a fuse to set off what was basically a firecracker. Once he thought he had harnessed the power of nitroglycerine, he began to cash in.

Invented in 1863, nitroglycerine was developed as an explosive through a U.S. patent granted to Nobel on October 24, 1865. Newton, writing 15 months later, and after reading some of the technical journals of chemistry, intended to at least give it a try. If it performed as efficiently and as safely as it seemed to be capable of, it would outweigh many startling newspaper reports. In the beginning, nitroglycerine demonstrated 10 to 13 times more power than gunpowder and using this new explosive could save more than $141,000 on the Hell Gate project.

Yet danger lay there. Nobel built a factory in 1866 in Krümmel; that same year the plant was destroyed by an explosion. Learned, as time passed, were rules to follow in transport, handling, and detonating nitroglycerine.

Balancing the virtues and dangers in his mind, Newton could not help but be impressed by the inestimable value of nitroglycerine. "Nitroglycerine required shallower holes, and fewer of them. Its smoke cleared faster than that of black powder, its debris cleaned up more easily, and it worked in wet rock, where powder did not."[26]

# FOUR

# Newton's Needs

## Planning and Progress

The clearest statement of what General Newton decided for the total project can be seen in the *Chief of Engineers Report* published in October 1867. It read:

> The project ... provided for the removal of Negro Head, Flood Rock, Hen and Chickens, Gridiron, Pot Rock, Frying Pan, Ways Reef, Shell Drake, the rock off Negro Point, the rocks near Woolsey's Bath House, Blackwell's Rock, portions of Hallet's Point and Scaly Rock, to a depth of 26 feet at mean low water.
>
> Sea-walls were designed also for some small part of the Middle Reef, Hog's Back, Bread and Cheese, and a beacon for Rhinelander's Reef. The estimated cost of the project was $8,692,645.15 [about $145 million in 2020].[1]

Federal spending in 1867 was posted at $584.8 million. Left out for the time being were Diamond Reef, Coenties Reef and the reef at North Brother's Island.

In less than a month (February 9), General Newton received approval of his extensive plans from the Chief of Engineers, who wrote "General Newton enters minutely into the method of conducting the work, the nature of the requisite machinery, and & derived from a close and careful study of the questions involved."

Still so much to study. Surveying would go on for many years, as would studies of the current's speed and the extent of the depths of the obstacles. The consequences of tides on Newton's plans also had to be explored since he would have to work on the river. Put simply, tides are influenced by the gravitational force of the moon. "The moon is strongest on the side of the Earth that happens to be facing the moon" and that strength draws water toward the moon. This National Oceanographic and Atmospheric Administration explanation also includes the force of the sun as it affects tides, as does the rotation of the earth on its axis and the movement of the earth orbiting the sun even as the moon orbits the earth. The interaction of the pull of all of these spheres causes tides.

Before too many years had passed, Newton had access to reports from the federal U.S. Coast and Geodetic Survey which explained

New York Harbor is visited by two derivations from the tide-wave of the ocean, one of which approaches by way of Long Island Sound, the other by the way of Sandy Hook Entrance. These two tides meet and cross or overlap each other at Hell Gate; and since they differ from each other in times and heights, they cause contrasts of water elevations between the Sound and the harbor, which call into existence the violent currents that traverse the East River … the meeting place of the tides to a space of about one hundred feet off Hallett's Point. Two tide waves visit New York Harbor, meeting and overlapping at Hell Gate.[2]

The General thought it best that the more extensive projects—Hallet's Point and Flood Rock—should be postponed until thrift and productivity might be brought to bear through advances in the technology of equipment and of the chemistry of explosives. Even after choosing nitroglycerine, the drilling of holes in the river rocks and reefs where the explosives could be inserted had to be solved. Newton knew also that the river at Hell Gate was at its least powerful from the turbulence of tides—and therefore the easiest and safest time to get work done—was during a period called "slack water," a period of only 30 minutes each day. How might the drilling be accomplished; how might the big project be finished if only working for half a hour each day? How might the divers be made safer and more efficient, being at the mercy of the swirling violent tidal currents? Speed in handling explosives, not caution, can often lead to disaster.

Before very long it was known that gunpowder needed 18 seconds to be completely consumed, whereas the nitroglycerine needed less than one hundredth of a second. One axiom for an explosion's power is that the speed of an explosion is a measure of its destructiveness. The detonation velocity of gunpowder remains in the range of 171 to 631 meters per second. The detonation velocity of nitroglycerine had been found to be 7,700 meters per second.

Measuring and testing and experimentation might go on in 1867 while New York City newspapers kept accounts of the toll on shipping. The schooner *Exchange*, sailing from Rondout, New York, to Providence, Rhode Island, with a load of Delaware & Hudson coal was wrecked at Hell Gate. The sloop *Vienna* traveling from Elizabethport, New Jersey, to Norwich, Connecticut, sank at Hell Gate on July 23, 1867. On May 12, 1867, the sloop *Rhode Island* after passing through Hell Gate, split her mainsail. On June 25, 1867, the schooner *E. Ewen* crashed into Little Mill Rock and was "entangled" on the rock. This stranding of the vessel necessitated the use of a pump to, in effect, lighten the ship and float it off the rock when it might be towed into port. Also towed off, the schooner *H.W. Bennett* on the Gridiron on July 16, 1867.

With all of these problems, obstacles and difficulties facing shipping, the Corps of Engineers in the person of General Newton planned on spending an amount equivalent to more than nine billion dollars in 2020. Simply, there was no faster or less problematic a passage into New York harbor for transatlantic ships nor for ships to and from New England and Canada, Hong Kong, Rio de Janeiro and San Francisco.

In addition to the demolition tasks facing General Newton was choosing his subordinates and contractors, and various forces in New York applied pressure. These included organizations of merchants who shipped out and accepted goods from all over the country and the world, organizations of ship owners, groups from insurance companies, along with organizations of politicians who were looking to find ways to increase their own power and influence and that the city as well. Overseeing Newton's work were federal legislative committees in Washington too, committees of commerce and national defense for example that work with cabinet level posts to see to it that transporting goods by ship—and building those ships—continue their importance to the commerce of the United States, while ensuring its safety from attack.

At least Newton had the advantage of easy movement within the city, certainly to get from his office in downtown New York to the Hell Gate sites. One way would be to board a horse car at Prince Street and travel north along what is now Fifth Avenue. Montague Davenport's book from 1876 described "The tram-cars … running in all directions, at a charge of six to eight cents, and are used by everyone."[3] Horses were used in the more densely populated areas of the city—that is, below 23rd Street—with steam engines used above.

Safety and public relations needed to be considered, as Newton later reported

> The proximity of [Hallet's] reef to habitations at Astoria, Ward's Island, and Blackwell's Island made it necessary to devise a system of explosion which, effecting the work of demolition, would at the same time do no damage to life and property. The atmosphere and the rock being the mediums through which the shock would be transmitted, it was essential that the waves propagated through these should be made as small as possible.[4]

The causes of ships put into danger at Hell Gate, being driven ashore or onto rocks, being sunk or needing to be towed into port, can be caused by the battle between tides and currents, as pointed out years before. If a ship's pilot intended to take the channel that is close to Astoria, he may find his vessel in danger if he gives Hog's Back Rock too wide a berth, and in so doing smashes into Mill Rock. Likewise, by steering too narrow a path the ship would smash into Hog's Back. And Hog's Back did not lie in the middle of Hell Gate, but to the north and east in the East River. Plainly said, the tidal currents of Hell Gate and its zone "bewilder the navigator and render the pass dangerous often to the most expert pilot."[5]

Yet Newton remained confident, looking into the future. "There is no doubt that a serious and well-considered attempt to push the work," Newton thought, "will be the means of stimulating mechanical inventions suited to this end, which, with the experience gained from day to day, will not only expedite the progress, but also materially diminish the cost, especially in the large item of removing the rock."[6]

## Hazards and Virtues of Nitroglycerine

Removing the rock would almost certainly entail the use of greater and greater quantities of nitroglycerine. What did the public know about nitroglycerine? Two years before Newton arrived, in September 1864, Alfred Nobel was notified not only of a major explosion at his nitroglycerine factory in Stockholm killing five people, most especially his younger brother, Emil. Not long after, the city of Stockholm enforced laws that experiments with explosives could not be made within the city limits.

The year before Newton arrived in New York, just a 13-minute walk from his office, an explosion of nitroglycerine was centered at the Hotel Wyoming at No. 333 Greenwich Street. Numbers that were reported varied but it seems that 18 people were injured and as many as 15 died, along with considerable property damage.

The year Newton arrived in New York, 1866, 70 crates of nitroglycerine exploded on April 3 while onboard a steamship in Aspinwall, Panama. Sixteen days later the true volatility of the liquid was demonstrated twice more. First, a leaking, unlabeled crate was taken to San Francisco's Wells Fargo office. Not knowing the contents of the crate, as workers forced it open, the nitroglycerine detonated, killing 15 people, some "blown to pieces and fragments of human remains were found scattered in many places."[7] Many more were wounded by flying window glass. Almost simultaneously, some of the nitroglycerine being used in the Central Pacific railroad construction detonated early and killed six people. Those accidents were enough for the California legislature to prohibit the transport of liquid nitroglycerine. If you wanted to use the explosive, you had to produce it very near to the project.

The year Newton submitted his proposal for Hell Gate, nine died from a nitroglycerine explosion in Bergen, New Jersey, just across the Hudson.

And yet, the substance might just be the only way to do the job the general was sent for. For a railroad's tunnel,

Engineer John Gilliss recalled, "In the headings of the Summit Tunnel the average daily progress with powder was 1.18 feet per day; with nitroglycerin, 1.82 feet, or over 54 percent additional progress. In the bottom of the Summit Tunnel, average daily progress with powder, full gangs, was 2.51 feet; with nitroglycerin, 4.38, or over 74 percent in favor of nitroglycerine."[8]

During the last quarter of 1867, passing through Hell Gate were 19,408 vessels, an average of 215 ships each day. Every delay, every ship that has to be pumped out or towed, every wreck, inflated the cost to the merchants. They must spend fees for piloting and fuel for the steamers, in addition to the salaries paid to the mariners. Wrecks or accidents meant that expenses were incurred for the replacement and repair costs of wrecked vessels as well as for the premiums paid to, and paid out by, insurance companies. By December the East River Association in the *New York Herald* claimed the "Number of vessels that go on shore" numbered as many as 25 per month.[9]

As in calendar year 1867, the bill appropriating the Hell Gate expenses passed in March 1868, but the money would not arrive to Newton's command until the new fiscal year had begun on July 1; $300,000 was appropriated for river and harbor improvements for the entire country during fiscal year July 1, 1868, through June 30, 1869, for prosecuting the work, reduced later on July 31, 1868, to $85,000 from the general appropriation for Newton's use in the East River.

Marking the true beginning of Newton's tasks, the money was to provide for "the removal to 26 feet depth below mean low water" of 12 rocks and reefs; "the construction of sea walls" on six others; and "the construction of a dike closing the channel between a rock just west of Hallet's Point called Bread and Cheese and Blackwell's Island."[10] It must be pointed out that estimates are just that. In the forthcoming years, there would be major downturns in the economy which would seriously affect funding for the projects.

Forecast by Newton as a ten-year project, the operations were aimed at seeing to it that a mean low-water depth of 25 to 26 feet in the river was reached for all rocks and reefs. This depth would improve navigation through the East River and Hell Gate well into the future. Much of the work's completion would depend a great deal on how much of each obstruction was under the water. Pot Rock, for example, extended out 130 feet into the channel, with 8 feet of water "upon it." Many of these rocks needed to be drilled and blasted to more than 14 feet to reach the required depth. At the beginning of Newton's work, surveys showed the current depth of the following: Pot Rock 20 feet; Frying Pan 11; Diamond Reef 17.5; Coenties Reef 14.3; and Heeltap 12.1.

Meanwhile, no money, though approved, was arriving for Newton's use even as the importance of the East River as a route through Long Island

Sound to and from Boston, Providence, Maine, and Nova Scotia continued, in addition to a route to the Atlantic as well. The *New York Herald*
claims that on January 30, 1870, alone "merchandise to the hundreds of millions of dollars passes through both ways and many thousands of lives" are
affected.[11]

The 1868 *History of New York* took notice not only of the goods that
left the port but of those that were produced there, writing "that this city,
by the census of 1800, returned a larger manufacturing product than any
other city in the Union, and more than any State, except New York, Massachusetts, and Pennsylvania." Stone also counted "the receipts for customs
in this port for 1865 were $101,772,905" (more than $2 trillion in 2020).[12]
(The pride and confidence in the city was also expressed in the words "Look
at our city now, in its extent, population, wealth, institutions, and connections," historian William Lette Stone wrote in 1868, "and consider how far
it is doing its great work, under God's providence, as the most conspicuous representative of the liberty of the nineteenth century in its hopes and
fears."[13])

On September 5, 1868, the schooner *Washington*, left South Amboy,
New Jersey, bound for New Bedford, Massachusetts. Sailing northeast from
its port in Raritan Bay would require a journey of about 220 miles eastward
on the open ocean. It seems to be the captain's choice to avoid the Narrows, that space in lower New York harbor through which he would have to
sail to get out into the Atlantic. The Narrows and Sandy Hook as has been
established, remained treacherous due to the accumulation of silt and sediment that formed sandbars and sandbanks. To avoid the problems of the
shallowness of the exit into Lower New York Harbor, the skipper chose to
sail northward through Arthur Kill and Kill Van Kull and so into Upper
New York Bay. From there he could sail his ship up the East River into Long
Island Sound.

Thus, the voyage would have the clear virtue of avoiding the open
ocean for as long as possible. The perilous condition of the Port of New
York is clearly shown by the fact that choosing the more sensible route did
not aid the schooner *Washington*. It sank in Hell Gate.

## *The East River Association*

The East River Improvement Association, formed in 1868 and sometimes attracting as many as 40 merchants, persisted in its efforts to attract
federal money. This group thought the East River must become the sole
entrance to New York harbor and called upon General Newton to speak to
this group even as the port of New York grew busier. In 1868, the Erie Canal

moved 3 million tons of freight, with most of it traveling down the Hudson River. In the harbor for one day's ship traffic, one could count 26 ships (barks, schooners and steamers) clearing the port of which 12 ships passed through Hell Gate going south; going northeast into Long Island Sound 48 ships passed through Hell Gate.

One needs to understand the equal or greater importance of shipping on the East River and the use of Hell Gate as a route. Pier numbering changes over the years. What were once "slips" or narrow spaces to tie up became piers. But in lower Manhattan some of these lower numbers became ferry docks. East River piers traditionally have ended at 90th Street, where the Manhattan side of the East River turns (ignoring Brooklyn for the moment). On the Hudson side the piers extend to about 59th Street, the current highest number being 99, again where there is a natural break in the flow of the topography. Then too, ships would not want to get too far from the Battery, perhaps for Customs purposes.

To accommodate these vessels' cargo and passengers, East River Piers numbered 4 through 68 extended from Broad to 62nd Street. Piers numbered 1–80 on the Hudson side ran from the Battery to 97th Street. Not appearing on "Bacon's Map of New York City" printed in 1868 were names of streets north of Houston, except for a few on the West Side from Sixth Avenue to Tenth Avenue. That same 1868 map shows a kind of wasteland (except for parks such as the one named after President Madison) until Central Park is reached. Where the Park ends at 110th, the map stops.

In its October 25, 1868, edition, *The New York Times* noted that each day $15 million worth of property passed through Hell Gate. Specifically, the East River Improvement Association calculated that, moving through Hell Gate in four months of 1868, 2,423 steamboats carrying an average cargo of 200 tons could be tallied. So, after a crash against the Hell Gate rocks and reefs, what was the cost to a crew, to an owner, and to the merchants eager to have their goods off-loaded or delivered elsewhere. The wreck of a vessel would block important sections of the river. Steamers might be sunk and their crews in peril; ships might be damaged but repairable; cargo might sink to the bottom of the river; vessels might be towed off the land or off the rock but be declared "unfit for service." New York merchants expressed envy for the stone piers in the harbors of London, Liverpool and Moscow, even as the Hell Gate passage carried on average more than $6 million worth of material. Unmentioned by the merchants were the fees for towing and dockage, piloting and dry docks.

From July 1, 1868, through June 30, 1869, Newton planned to begin with obstructions nearer the southern tip of Manhattan than the more northerly Hell Gate. Battery Reef (really a shoal formed by deposits from currents); Diamond Reef (between Governor's Island and the Battery, reaching 366 feet long and 255 feet wide), and Coenties Reef (600 feet from Coenties Slip pier in downtown Manhattan on the East River side, extending 250 feet long and 130 feet wide). The amount appropriated during the fiscal year reached $180,000. The cost would be $463,000 to bring those chunks of gneiss to a depth of 24 to 26 feet. He also oversaw surveys of various bays and inlets on the western end of Long Island from July 1868 to February 11, 1879.

For work on the rocks, the bidding was won in September by Sidney Shelbourne, signing a contract to last until December and later extended until August 1869, mostly working with experiments on drills. It remained certain that the Maillefert method did not work. Perhaps Shelbourne's drills might.

When not conducting surveys and talking to interested parties, General Newton was known to have concluded that "For the removal of Hallet's Point Reef it was determined to employ a process of undermining the rock by tunnels and galleries, from which mines should be exploded to break up the whole mass of the rock at once."[14] (The use of "mines" was in the modern sense—an explosive such as a minefield or floating naval mine.) His plan seemed bold at first glance but, as was typical of Newton, he knew of three major proven precedents.

The shaft and tunneling method had been worked on by C.G. Reitheimer, a scheme adopted by Newton. Reitheimer's work was performed at a Hollyhead, England, quarry in January 1867 with gunpowder as the explosive.

Next, closest to Newton's time and place, is found in the work of United States Corps of Engineers General Alexander and San Francisco civil engineer Allexey W. Von Schmidt, for the removal of Blossom Rock, in San Francisco Harbor on October 3, 1868. This work was supervised by two more men from the Corps of Engineers, R.S. Williamson, and W.H. Heuer, and their report to the chief of engineers would soon be read by General Newton. (In fact, Newton would later request the services of Heuer during the 1872 fiscal year.) Blossom Rock was, however, smaller than any one of the East River rocks. The San Francisco Bay rock measured 180 by 100 feet in area, while Way's Reef was surveyed at 235 by 195. (In square feet, Blossom Rock covered 45,825 square feet; the Hallet's Point project measured 130,680 square feet.) This project too used gunpowder as the explosive.

Then slightly later in 1869, a 23-page publication appeared called *Submarine Blasting in Boston Harbor, Removal of Tower and Corwin Rocks* by Lieutenant Colonel John G. Foster, also in the Corps of Engineers, whose manual on underwater demolition was relied on for 50 years. Up until then, the art of blasting under water was almost unknown. Even so, the word "nitroglycerine" appears nowhere in the manual.

Newton's Hallet's Point project would not come to fruition until 1876, even while he acknowledged his work on the river from the beginning in 1867 "the particular machines described by me were not relied upon as the sole, or even the best, means of effecting the object desired ... and to furnish data for its estimate. One thing became certain, that the boring should be conducted from a platform above water."[15] But no such platform existed—Maillefert worked from rowboats—and so Newton continues "I have been considering other means of removing the rock from the bottom, after blasting."

But blasting remained a problem. Two and a half years earlier the *New York Times* talked about how "the public were startled by reports from different quarters of a succession of extraordinary explosions. From Aspinwall, from Sidney, New South Wales, from San Francisco, there came one fearful story after another of lives sacrificed by the hundreds.... Nitroglycerine, the infernal compound which ... was known pretty generally to science; but as an article of commerce it was scarcely known at all."[16]

The *Titusville Morning Herald* reported on an "Explosion at oil exploration site in Reno, Pennsylvania." The nitroglycerine detonation resulted in a death and "About fifteen houses of all sizes including the Reno Company's building, were more or less sprung and shattered."[17] Three months later, a headline in the *New York Times* said in Urbana, Ohio, a locomotive and four cars were destroyed and "a house a quarter mile from the explosion was demolished by the concussion" of the nitroglycerine. An "iron rail track was thrown three hundred yards ... and bent double."[18]

In the handling and careful use of nitroglycerin (or any explosive and device), Newton was fortunate to have the school—at what was officially named the Fort at Willets Point—working on the substance. Major Henry L. Abbot contributed greatly to Newton's projects. Appointed on June 1, 1866, to command the Engineer Battalion and Engineer School at Willets Point, a kind of graduate school for the officers trained at West Point in engineering, Abbot set up a laboratory to investigate "the laws of translation of destructive shocks of explosions under water."[19] The focus for these investigations was, as might be expected, the destruction of naval vessels. It was not unusual for him to conduct several hundred trials for one explosive. Willet's Point lies in Flushing Bay south of the East River, north of Hell Gate, and the site is the same as Fort Totten, just across the river at Throgs Neck. Included among Abbot's many publications are *Approved Matériel of the U.S. Defensive Torpedo System and Submarine Mining, in the Defence of Harbors and Rivers*.

On August 21, 1868, Newton put out bids for some of the Hell Gate projects.

## At Work with Sidney Shelbourne and Benjamin S.H. Maillefert Again

The private contractors who had been hired to demolish the smaller obstructions Way's Reef and Frying Pan Rock continued to have problems. Sidney Shelbourne began work on January 12, 1869. But within four days the *New York Herald* saw that "Mr. Shelbourne's operations at Hell Gate have as yet been simply preparatory. It was found impossible to make any anchor or weight to hold the tug in place during flood tide. The tides and currents remained as treacherous and powerful as ever. The plan finally adopted was that of drilling holes in the rock, in which anchors are to be fixed, and these holes have been made."[20] Certainly, the problem of stability in the river remained unsolved, if even a steam-powered tugboat could not remain stable in the river. And the *New York Times* was quick to point out in 1869:

> The necessity of improving New-York Harbor—its docks, wharves, and channels—is becoming more and more obvious and imperative every day. The concentration of numerous and important inland trade lines at this point has already made it the maritime center of the continent and with the new road to the Pacific opening a remote Eastern traffic. It must soon rival in importance the largest ports of the Old World.

Frying Pan Rock was located halfway between Ward's Island and the village of Astoria. The smaller rock (12,475 square feet) can be viewed as a part of the chain of rocks in the Middle Ground. Shelbourne hoped to be able to sink a steam engine and drill encased in a water-tight chamber to the bottom of the rock, drill a number of holes in the rock, and raise the engine. The holes being made ready, divers would be sent down to place the explosives in the holes. No one seemed to be placing the drill. Once the divers had been moved away from the blast site, the explosives could be detonated and another contractor would be hired to grapple the debris, bring it up to the surface, and remove it via a scow. It was concluded by *America Illustrated* in 1889 that "Shelbourne, who proposed to work by drilling and exploding the rocks, did little more than make various experiments and demonstrate, at the cost to himself of twenty thousand dollars, the impossibility of certain modes of operation, before the time specified in his contract expired, and the contract was withdrawn."[21] One of the modes was an attempt to work with a machine 35 feet high, 27 feet in diameter, and weighing 28 tons, a machine which was driven at first by water and afterward by steam power. This machine was finally wrecked by a colliding scow.

The contract with Mr. Shelbourne, extended to August 15, 1869, expired with no progress "except the constructing of machinery for boring and removing them, and up to the date of limitation no practical test of the machinery had been made, although the apparatus had been located on the Frying Pan Rock for that purpose, but was collided with by a tow passing through the Gate, and left a wreck upon the rocks."[22]

Beginning around August 18, 1869, three days after Shelbourne, Maillefert's firm commenced with operations on Way's Reef by the method known as "surface blasting." Maillefert's crew's system, not any different from his methods more than 15 years before, was to lower a can of explosives with sandbags attached for weight. With wires connected from the can to a battery on a rowboat, that small boat moved 100 feet away from where it had sunk the can to set off the explosion. Although it became clear that this method produced little results, Maillefert kept believing. Nevertheless, so little was accomplished that the *New York Times* complained in September "We are inclined to think it about time that M. Maillefert gave us something more substantial than rose-colored newspaper reports of his progress in blowing up … Way's Reef."[23]

The complexity of the work required by Newton's orders is underlined by Shelbourne's problems and by the fact that Newton admits in the *New-York Tribune* that "Hell Gate's bottom must still be thoroughly mapped, and that charting will probably take two years. Mr. Shelbourne's firm will be hired to do grappling work on the river."[24]

Clearly, gunpowder was not the answer. Nitroglycerine might be. But the country's newspaper readers and Newton could be warned again, this time by four reports of horrible accidents: on the building of the Lebanon Springs Railroad; in Hudson City, New Jersey, and two in Titusville, Pennsylvania, all involving nitroglycerine. The count in 1869 stood at four dead, eight injured and a building blown up.

Of these explosions, the scientist who put to use the nitroglycerine later explained the number of stupid ways that his oily liquid had been misused. Nobel says in *Scientific American*:

> In five instances, congealed nitroglycerin had been melted purposely over the fire. In three instances a red-hot poker has been plunged into the oil in order to melt it. In one instance a man took to greasing the wheels of his wagon with nitroglycerine, knowing what it was; and it went all right until it (the wheel) struck hard against something, and then the whole went to pieces. In one instance it was burned in a lamp as an improvement on petroleum.[25]

In 1866 Nobel had traveled to the United States, and successfully defended his patent rights there. Having assigned his patent to the United States Blasting-Oil Company, Nobel saw his bank account increase by $270,000. He also was a partner in the Atlantic Giant Powder Company. His new company built a factory to manufacture nitroglycerine in New

Jersey, not far from Fort Lee in 1866. It was destroyed by an explosion in 1869.

During the spring of 1869, an additional federal allotment of $178,200 was granted (up from $85,000 from the 1868–69 amount) for the East River project from the general appropriation for the Improvement of Rivers and Harbors. With friends still in the city supporting Mr. Maillefert, a contract was then made with his company for the removal of Pot Rock, Way's Reef, and Shelldrake, at the rate of $44.28 per cubic yard. (To allow some perspective, one year later with Newton in charge the price dropped to $5.75 per cubic yard.) Maillefert was the master of gunpowder. But at the end of the Professor's contract, May 5, 1870, that era of gunpowder's use for blasting large hard rocks was over.

Newton's estimates served as requests for money for a single fiscal year but, as expected, the amount of money allocated fell far short. Others than Newton were displeased. On May 27, 1869, the East River Association planned to travel to Washington with Newton's estimates and go before the House Rivers and Harbors Committee to ask for $2,000,000. The money from the federal government had been arriving fragmentarily, and much of that money necessarily being used for surveys, tide measurements, ship traffic reports, and the like. Itemized as well was a list of needed refurbishing of more than 60 working piers and docks on the East River alone.

The General's 1867 estimate for labor, still making its way through Congress, was part of the cost.

| Two clerks, at $130 each | $3,120 per annum |
| Two draughtsmen, at $130 each | $3,120 per annum |
| One messenger, at $60 | $720 per annum |
| Two chief divers, at $380 each | $7,600 for ten months |
| One master mason, at $150 | $1,800 ten months |
| One overseer of laborers, at $120 | $1,440 for ten months |
| One master blacksmith, at $120 | $1,440 for ten months |
| One master carpenter, at $120 | $1,440 for ten months |
| One rigger, at $100 | $1,200 for ten months |
| One storekeeper, at $120 | $1,440 for ten months |
| Four sub-overseers, at $60 each | $2,880 for ten months |
| Six night watchmen, at $50 each | $3,600 for ten months |
| Forty laborers | $18,960 for ten months |

While the budget request slowly went into committee meetings, workers on East River (Brooklyn) Bridge works began June 1869.

There remained some opinions in the city that all of the obstructions in the river cannot yet be known. It must be pointed out that many things in the river could be measured: the timing of tidal movement, the flow speed of the current, the depth of the river. But the unpredictability of eddies, whirlpools, winds, and snow had to be taken into account along with possible collisions with working boats on the river itself if Newton were to attack the rocks with any efficiency of time and of cost.

## The Case of the Thatcher Magoun

No matter the randomness on the East River, particularly at Hell Gate, Sandy Hook's entrance to the harbor remained a problem. For instance, a clipper named *Thatcher Magoun* sped from San Francisco in 96 days having successfully mastered the winter on the Atlantic.

But when it arrived at the east bank of Sandy Hook and put into the hands of a pilot to maneuver the ship safely into Manhattan, the ship grounded, forcing lighters to come to the ship and offload its cargo. Then, without damage, it had to be pulled off the sandbar by two steamers. The ship survived the hundreds of miles from San Francisco, but on its one hundredth day of transporting cargo, it cost the owners even more fees for a pilot, for lighters, and for steamers to act as tugs.

On the Hudson River, then called the North River, Newton took on responsibilities as head of the Corps of Engineers for New York City. Projects numbering ten for various repairs, for maintenance or demolition from south of Albany down to Castleton-on-Hudson, involving dikes, channels and sections of islands came under his authority. On the East River, near the tip of Manhattan, he also tended to the improvement of Wallabout Channel.

Many projects took up his time, limiting his funds for Hell Gate. Clearly, too, federal money had to be spent on all the other projects in the state, with that one section of one river not drawing much attention or funding. The reality that, for the federal government, the older Committee on Roads and Canals had just been changed to the Committee on Railways and Canals demonstrated not only a recognition of the growing importance of rail transportation but, for Newton, perhaps an additional problem in funding. The fact that railroads in the United States neared 52,000 miles, doubling the miles of ten years before could not be ignored.

On May 10, 1869, the transcontinental railroad was completed. At the same time, the famous ship builder and America's first true naval architect,

William Webb, constructed his last ship—the last packet, and the last full rigged ship, built in New York—the *Charles H. Marshall* in 1869.

These beautiful vessels would sail on, particularly on lakes and rivers for many more years, but the voluminous holds of steamers were the future.

Newton had to answer to many people including local politicians along the Hudson. His work also would be inspected by the Commissioners of the Dock Department for the city. What might impress all those with a tight grip on the purse strings was Newton's new plan for Hell Gate, which appeared on June 9, 1869.

General Newton proposed a new approach to one major problem: the three acre reef that projected out into the river from Hallet's Point, on the northeast portion of current day Astoria. The southernmost section of the Point was easily located by finding the remains of old Fort Stevens, a small fortification for just 12 cannons built in 1814 during the War of 1812. The fort was apparently abandoned after 1815, having served as a temporary structure.

The General's plan required tunnels radiating out from Astoria and then under the river itself. According to this newer plan, the work will be completed more quickly and more thoroughly. It would exchange surface blasting with small blasts and drilling to make tunnels.

As for the rocks in mid-river, Newton published a proposal, writing to his superior, Maj. Gen. A.A. Humphreys, Chief of Engineers. In a report on June 9, 1869, Newton outlined "a contingency which at least is possible—to construct a boat and machine for the government ... to stand the shock of collisions."[26] The craft would have to be mobile, using tugs to move it up and down the river when needed, but remaining stable when in place at a rock or reef. This proposal would become a reality in July of 1871 when a vessel designed by Newton himself was completed. General Newton called it a steam drilling cupola scow.

Meanwhile, another voice was heard by Newton on October 29, 1869. W.W. Vanderbilt, constructing engineer of the Pacific Mail Steamship Company, proposed a 2,000-foot-long, 600-foot-wide canal through Astoria at a cost of $3 million paid to private contractors, but financed by the government. (Such a canal had been proposed as far back as 1829.) In the *New York Herald* of January 30, 1870, Newton's perspective was published from the general's letter to the Secretary of War, Newton disagreed with the canal plan, writing that the Corps of Engineers needed to work on Hell Gate, that way "millions are saved ... by prosecuting the work itself instead of letting it to a contractor." Furthermore,

> It is doubtful whether such a canal would connect directly with the channel east of Blackwell's Island. It is doubtful that such a canal would be of any use to transoceanic commerce ... for it would close up to maritime trade the entire

east shore of New York north of Eighty-Sixth Street, where European commerce is destined to house itself sooner or later after the clearing out of Hell Gate.

## Work Begins on the Reef at Hallet's Point

Newspapers heralded the beginning of a complicated, even revolutionary, project on the East River on August 12, 1869, by pointing out the peculiar danger of that reef. "In avoiding the point, ships run into Flood Rock. Hallet's is a singular problem since it is not only a projection into the channel, but also it produces eddies on both sides of it depending on the flow of the tides."[27]

Hallet's Point extended 360 feet into the East River. Newton was later quoted as saying for "the first time … that tunneling was suggested as a way of removing rocky obstructions in a channel."[28] Here he meant demolition in a narrow space between populated areas. While the work at the Point might be begun, three reefs in mid river—Way's Reef, Pot Rock, and Sheldrake Reef—would have to be attacked as well.

At Hallet's Point, Newton now knew what had to be done. The job site at Hallet's Point was found in Astoria, Queens, approximately opposite 92nd Street in Manhattan. Each day at work Newton could watch the sailings of the Astoria ferry which, on the Queens side, later featured trolley tracks and retail shops and a hotel near the docks.

The ferry embarks from Manhattan's 92nd Street and docks on the Queens side in Hallet's Cove today, just south of Astoria Park and west of 21st Street. (Its usefulness to the city did not end temporarily until 1936. Service to Astoria from Wall Street began again in 2018.) From that Astoria dock, the steamboat *Morrisania* ferried commuters to Harlem.

Though the hard rock dagger that extended from Hallet's Point only the three acre portion of the point, the reef, itself would have to demolished, severed, to ensure safer passage for the many vessels using the East River. One method Newton decided on was what might be called "mining," small scale shaft mining, which is to say the making of tunnels in rocks.

Partly it was a mere 30 minutes between the incoming power of the two tides that made some of the work difficult. For more than 140 years, this description by General Newton of Hell Gate has been repeated:

At a time when other waters settle into slack, the downcoming tide, which has been delayed four hours by the distance and the drag of the Long Island Basin, begins its relentless drive—and the struggle for mastery is on. Four hours after entering the Sound, this tide has changed the flow of the river which is down the narrow "sluiceway" from the Bronx and down Little Hell Gate channel into Hell Gate Basin, counterclockwise around Mill Rock and as far down the river as the upcoming tide will allow. To this confusion of ebbs, add the rocks, reefs, and freakish whims of the winds. At ebb tide, the process was reversed, but no less confusing.[29]

To counter the tides and currents around the projection in Hell Gate before Hallet's shaft and tunnels could be dug, a coffer dam was needed to protect the entire project. This type of dam, whose history goes back almost 2,000 years, was intended to protect the space of land and its tunnels being worked on by preventing water from entering the workspace. Once protected, buildings for carpenters and blacksmiths could be built as well as sleeping quarters for watchmen and others. A central storage shed would be built as well behind the dam. The wooden wall to surround the Hallet's Point projection had been slated to begin in October of 1868.

The coffer dam was not begun until midsummer 1869 but was quickly pronounced complete on October 20, 1869. In that same month, Newton could begin the excavation of a shaft, initially removing 4,841 cubic yards of rock. To match the shape of the dam, the shaft's configuration had to reach 32 feet in depth. Newton thought it appears as "a polygon, with a circumference of 443 feet, with a mean interior diameter of 100 feet."[30]

Once the proper depth was reached for the shaft, then a network of galleries (tunnels) was carved under and into Hallet's Point but at the same time leaving pillars for supporting the roofs of the tunnels (also labeled "galleries" in reports). This work was barely begun in July 1969, five months before the shaft at Blossom Rock in San Francisco, using the same sort of general approach.

The drawing on the opposite page gives a broad view of the site and even includes visitors strolling along the top of the dam. The names of the tunnels can be seen on oval signs above the opening of each. When finished, "an examination of the dam showed it to be built of heavy timber and securely fastened to the rocks by bolts."[31]

*The New York Times* spent some words carefully listing the progress on a few rocks that M. Maillefert claimed. The Frenchman excused himself from not having accomplished more by blaming the government for not providing him with sufficient charts. He asked that his contract be extended. At Hallet's Point, the *Times* reported in November, "the mining goes on slowly owing to the lack of necessary machinery for hoisting, etc."[32] The work presented another problem. It seems that the upper layer of rock was easier to shatter, but as the depth increased the rock was found to be much more difficult to break. The difficulty might be solved "When General Newton's soundings have been made [and] there will be more accurate data to judge from."[33] Newton's office was later to write how little the French engineer had accomplished on the river and his contract ended in 1870.

That same November the sinking of the shaft began on November 1 and continued until the middle of June 1870, when operations were

A view of the coffer dam holding back the East River and the entrances to the tunnels that had just begun to be drilled and blasted under the river (*Scribner's Monthly*, November 1871).

suspended temporarily on account of the funds available for this part of the works being nearly exhausted. By that June date, "4,841 cubic yards of rock were removed, at a cost of about $5.75 per cubic yard. A cubic yard of rock is usually the equivalent of 2,700 pounds; thus, the workers in the shaft had excavated 13,086,900 pounds of rock."[34]

To excavate the rocks, some explosives could augment picks, shovels, hand drills and the like. Help would come from the 1866 establishment of a kind of graduate school for Army engineers at Willet's Point where experiments of science with "submarine mining," the army's use of explosives underwater, were about to take place. Major Henry Larcom Abbot, the chief at Willet's Point, saw that harbors could be protected from attack if enemy vessels could be blown up if they came in contact with the explosive mines. U.S. Army engineers conducted experiments and studied the work of other engineers, analyzing the studies in publications from all over the world. Experiments with "torpedoes"—now called mines—were also conducted in the River itself. What was learned about the nation's defense would come to be useful in the demolition work on the East River.

# FIVE

# Newton Adapts

## *The First Steps*

The decade of the 1870s saw great changes in technology. The 130 feet high Equitable Building in the city featured passenger elevators. The decade would see advances in drills, the engines that powered the drills, and the drill bits as well. These, and much more, would become vital to Newton's work being possible. Newton's willingness to adapt to the more efficient tools would matter no less.

The size of the project, and its cost, might be great but Newton knew "as each reef that is removed helps greatly to reduce the current and so to simplify navigation as to enable pilots to keep control of their vessels and avoid collisions, the uncertainty of doing economical work from the surface in Hell Gate is rapidly disappearing."

Once money from the federal government resumed after July 11, 1870, the digging of the shaft could continue. Beginning at the foot of the coffer dam, "a sufficient number of galleries" would be started "under the reef, to enable charges of explosives to be so placed as to break up the whole reef at one blast."[1] Eventually, work on the reef meant removing 137,700,000 pounds of rock.

With a 32 foot shaft complete, Newton had to establish an entire workshop at the shaft's bottom. Sitting in his office at the Army Building, at Houston and Greene streets in the fall of 1870, Newton wrote

> Adjacent to the dam were erected a smith's shop, a store house, and a boiler house to aid the progress of the work. A small brick magazine for storing powder was put up on that small above-water part of Flood Rock, a short distance from the scene of operations. The broken rock was removed from the excavation by means of a derrick operated by a steam hoisting engine, and being put on trucks was dumped on the marsh adjacent to the works.[2]

This was still dangerous work with laborers working into the rock as well as under the river.

This method of excavation had been proposed by General Alexander,

This view of the Hallet's Reef work area demonstrates its relative primitiveness. The entrance by a long, thin ladder and the presence of workers using pick axes provide clues to the crude material Newton began his work with (*Harper's Weekly*, September 23, 1871).

of the United States Corps of Engineers, for the removal of Blossom Rock, San Francisco Harbor, but that reef was a very small one in comparison with the two large reefs to be removed at Hell Gate. Like Blossom Rock, the central idea is that the reef "will be lifted in the air and shivered to pieces."[3] "Shivered" might seem an odd word to describe the breaking up of rock. Nevertheless, that is what happens in an explosion because it is a heated shock wave that breaks up the rock. One of the measures of an explosive is called its *impact* sensitivity. Gunpowder will explode if a weight is dropped 16 inches onto the material. For nitroglycerine, it is 1 inch.

The work under General Newton went on slowly but as *Harper's* new monthly magazine pointed out in July 1870, "the commerce of North America flows through New York; and since ... the traveler who wishes to proceed from almost any point on the North American continent to any other part of the world, in order to go conveniently and comfortably, must go through the New York door."[4] For example, the shorter route out of New York harbor, the route via the East River and Hell Gate, meant for some portion of the journey the smoothness of the trip out through Long Island Sound could last more than 100 miles, as far as Newfoundland.

## Ships and Vanderbilt's Trains

By 1870, shipbuilders in New York had drastically decreased in number. The initial cost of building iron steamships was simply too expensive. The ship building yards were equipped with tools for wooden ships; retooling would be prohibitive. The same issue of *Harper's* also acknowledged that already "There are thirty or forty of these huge [steamship] structures … that … connect it directly or indirectly with every important commercial point on the globe…. Sometimes eight or ten of these steamers leave the port on the same day."

There were three types of vessels arriving in New York: regular passengers, freights, and emigrants. The United States could count just 196 of its own steamers, while its sailing ships built years before totaled ten times that at 1,946. Steamers from other countries arriving in the city's harbor equaled more than 500 vessels. And on the Hudson, the appearance of propeller-driven American steamboats, aided by tugs, meant that as many as 125 barges could be towed, with the tugboats nudging a few barges off at river ports and picking up some as well. Nevertheless, in 1870, sail ships still accounted for 85 percent of the total maritime tonnage.

One way of seeing the changes in the city was through transportation. One ferry line owner, Cornelius Vanderbilt, would affect transportation in New York by switching from being a boat owner to a railroad buyer. By acquiring both the New York & Harlem Railroad and the Hudson River Railroad, Vanderbilt had bought use (but not ownership) of Albany's newer and stronger 1866 Hudson River Bridge (now called the Livingston Avenue Bridge). Its status grew as the only rail bridge across the Hudson River and for five years the only bridge that could accommodate railroad traffic to cross the Hudson.

This meant Vanderbilt now owned a rail line from New York to Buffalo, a route once dominated solely by the Erie Canal. Once he leased the use of the bridge to the Boston & Albany Railroad, Vanderbilt could offer the speed of rail (about 40 MPH) to passengers and freight to travel from New York City's Chambers Street to Boston (320 miles). Other transportation companies took customers by steamer up to Connecticut where they could then board a train for Boston. And even then, boarding a ferry or two might be needed to complete the trip.

So companies like the Fall River Line moved steamboat passengers to where railroads had a terminus, saving about 100 miles from the trip by rail, the trip lasting 13 hours. Steamers moved between 10 and 15 miles per hour, while trains sped at 20 to 25 miles per hour. With the increased speed, with the railroads not impeded by frozen rivers, and with their ability to offer travel in comfort, railroads soon grew rapidly. In 1830, there were 23 miles of railroad in the United States. By 1840 this had grown to 2,808, and by 1860 there were 30,626 miles. Railroad miles grew from 1860 to 1870 to almost 53,000 miles.[5]

New York State recognized these coming changes as early as 1840 with its State Committee on Railroads. Money previously invested in shipping moved to investments in railroads and funds were found for buying the land, laying the rails, and buying the rolling stock, the locomotives, and cars. Travel remained limited by the simple fact that railroads in major cities connected only with other major cities. For example, the Northern Pacific Railroad could carry goods from Minnesota to Seattle but there were few connections to get to Minnesota by rail.

## Dockage and Piers

And while railroads began to reach out, it is clear from engravings and photographs of the time that many of the ships docked in New York were headed for ports in the South, where railroads were less extensive. Taking passage to New Bern, North Carolina, population under 6,000, could be found right in New York harbor. So while the North moved apace on building rail lines, the South continued to stress their ports. Shipping may have been diminishing but certainly remained vital.

To move just within the metropolitan area, for so it might be called with a mostly Manhattan population of 942,292 (an 8 percent rise from 1860), ferries came to the fore. By 1870, the Union Ferry Company of Brooklyn had purchased five ferry lines, and by 1873, passengers could choose from 28 different ferry journeys. Two of them could take you right into Hallet's Cove in Astoria to view the work at Hell Gate: The 92nd Street Ferry and one, from Peck's Slip at the bottom of Ferry Street, a street that no longer exists.

Shipping did not slow down. Arrivals and departures at New York just for coastal traffic grew from 14,370 in 1865 to 28,665 in 1870 and many of those ships were much larger steamships than the 1860 types, and certainly with larger capacity than the clippers.

It was thought that the facilities for docking in New York City lagged behind those in Great Britain. The terms used for places where ships may remain stationary afloat but attached to a structure are these: wharves, piers, docks, basins, and slips. Each of these words refers to a site where goods may be transferred to land and passengers may board and disembark. That New York City waterfront for commercial purposes extended for 27 miles from 51st Street on the East River to 61st Street on the Hudson River. In response to the call for changes, on March 29, 1870, the

Department of Docks was created by the New York City Charter, which appointed former Civil War General George McClellan as chief engineer of the new department. Some bold measures were called for.

## Despite Nitroglycerine's Deadly Accidents Newton Trusts It

Newton's early reliance on nitroglycerine, however, might startle New York City's newspaper readers because of a series of the seven reports of disasters from the explosive in 1870. It seemed that the nitroglycerine's problem was that whenever it was transported "the various shakes and exposure to [warm] weather" caused the chemicals to explode. It was thought that the shipping problem was solved when George Mowbray built a factory for making the explosive at the site of the Hoosac tunnel excavation in 1867.

There, eventually were built seven buildings, the main being called the Acid House "because of the chemicals and troughs used in making nitro."

In his 1978 tunnel history *A Pinprick of Light: The Troy and Greenfield Railroad and Its Hoosac Tunnel*, Carl R. Byron explained:

> [T]he Acid House housed soapstone troughs in which the acid mixture was prepared. This mixture was carried into the hundred-foot-long "converting room," where 116 stone pitchers sat in nine ice-water-filled troughs. Pure glycerine from glass jars shelved above the troughs dripped slowly into the acid-filled stone pitchers while fans blew the fumes generated in the reaction away. If the fumes built up because of a fan failure, a massive explosion would result.[6]

Yet in the very first month of 1870 not everyone was knowledgeable or careful, as Mowbray's works seemed to be.

1. January 4, 1870—an explosion on a wharf in San Francisco occurred after a boy sat on a leaking can of nitroglycerine. His mother pulled the boy up, and swatted his behind, the force of which in the air set off the explosive, "whereby the woman, her ill-fated son, and their house was shattered to atoms."

2. "Nitroglycerine Slaughter," Friday, March 18, 1870—Explosion of the nitroglycerine factory, Ridgefield, New Jersey—four dead.

3. "Nitroglycerine Explosion," Saturday, April 9, 1870, two were killed, "their bodies thrown a great distance … and were horribly mangled" at Ridgefield, New Jersey.

4. "Another Nitroglycerine Accident," Saturday, May 7, 1870—the west end of the Midland Railroad Tunnel, at Wurtsboro, Sullivan County, New York—"A Man Blown to Atoms … blown up into the air 300 feet and 50 feet away."

5. "Nitroglycerine Again," Friday, June 24, 1870—one killed, 30 wounded at Worcester; three freight cars demolished.

6. "The West," Wednesday, November 2, 1870, Painesville, Ohio— 150,000 pounds of nitroglycerine exploded; "four persons were blown to atoms."

A seventh event was the most shocking of all, since it happened to the man who was chosen to be the Hoosac Tunnel project's superintendent, John V. Velcher. Velcher was trying to safely carry away a tube of fulminate of mercury from the "acid house," or place where the nitroglycerine was stored, when the tube exploded and that caused the acid house to be demolished. As to Velcher, "about twenty pounds of his remains have been gathered up, placed in a casket, and deposited in the tomb."

The eighth and final explosion had wide repercussions. Nitroglycerine as an explosive was a new and, to many, frightening substance. The citizens of New Jersey were fearful of a factory in their state and *The Plantation* had noted in its April 9 issue, "the residents had procured the passage of a bill in the New Jersey Legislature, ordering the removal of the dangerous manufactory" of nitroglycerine. But the bill's effect had not taken place before a "nitroglycerine factory ... on the Hackensack River ... was blown to pieces, four men killed, and a number of others severely wounded. ... Bricks and mortar were hurled in every direction. ... The bodies of the unfortunate workmen were horribly mangled."[7]

General Newton continued to study his projects in September 1870. Looking west from Hallet's Point across the river, Newton knew that there lay a line of dangerous rocks and reefs which can be thought of as a straight line towards about 92nd Street in Manhattan. The four obstacles between those two points were named Flood Rock (of which The Gridiron was a part), and The Middle Reef which included the two Negro Heads as well as Hen and Chickens. Those being destroyed, a wide passageway would open up through Hell Gate. All needed to be attacked and in what order? Should they be demolished from above or from below? How may this work be accomplished?

Even so, Newton now had a clearer picture of the work ahead at Hell Gate and elsewhere through surveys, studies of the tides and currents, and figuring in the increased traffic and increases in sizes for newer steamships. As he waited for his annual proposal to move through Congress, Newton remained active supervising many projects, work that included removal of the wreck of the steamer *Scotland* in New York Harbor.

This wrecked vessel could serve as a model experiment for submarine blasting for Newton. Later it was brilliantly stated,

Breaking effect under water—Five pounds placed in a stone jar, and suspended against the iron side of the steamer *Scotland*, sunken off Sandy Hook, cut, as with a knife, a fissure ten or twelve feet in length. It is believed that an equivalent amount of powder, similarly placed, would have rocked or swayed the whole side, but would not have cut it; first, because its action would not have been percussive; and next, because its initial, or rather resultant force would have been spread over a larger area than would that of the equivalent of nitroglycerine, the less bulky material.[8]

## Newton's Confidence

But Newton had non-explosives work as well, such as strengthening fortifications on Staten Island, and Fort Montgomery (50 miles north up the Hudson), as well as more distant work improving Burlington Harbor, Vermont, and Plattsburg Harbor, New York, and the Appomattox River in Virginia. Far downriver the first signs of the bridge to Brooklyn could be viewed. (The man who had started that project was Washington Roebling, yet one more [former] member of the U.S. Army Corps of Engineers, serving as an engineer at the battle of Gettysburg.) That project too, setting its bases at the East River's edges had to deal, as Newton did, with the tidal wave flowing into New York harbor which met the downflow of the "Hudson River at the Battery where it is forced up the East River and through Hell Gate expending its force in filling the bays of Long Island Sound."

The work at Hallet's drew increasing interest; a description of a visit: "At Hallett's Point … we were delivered over to the hands of the superintendent, Mr. Reitheimer, who entertained us very pleasantly, and showed great politeness to the ladies and gentlemen of the party, especially to the ladies. He explained all about the works, and … showed us his plans and specifications, and then induced us to step into a wooden box slung at the end of a derrick, and be lowered away into a pit of fifty or sixty feet in depth."[9]

Knox would go on to describe that at "Hallett's Point … the reef extends more than three hundred feet from one shore so that the actual width of the channel is reduced to three hundred feet. The water boils furiously over this reef and turns a large part of the tide upon the Gridiron Rock, frequently throwing ships upon it."[10]

In the East River, work continued on Frying Pan Rock (halfway between Hallet's Point and Ward's Island) and Shelldrake Rock (in Pot Cove) as well as possible while trying to deal with 8 MPH currents. While the seven-year Hallet's Point project took shape, demolition up and down the river went on, particularly at this period on Ways Reef (near Pot Cove).

No matter what Newton was doing, no matter the breadth and originality of the plans he had revealed, in the opinion of *Scribner's* in 1871, "there were few engineers who did *not* seriously doubt the possibility of dislodging by human agencies those solid barriers which had so long-withstood the terrific force of the tides sweeping over and around them, and which were accessible only for a few minutes at slack water each day."[11]

Newton, deaf to these claims of impossibility, commenced the practical work in May 1871, upon Diamond Reef, south of Hell Gate near the mouth of the East River. By September 1871, the scow was found at work on Coenties Reef, Shelldrake, Way's, Hog's Back, and Pot Rock. Newton had estimated more than four years to be the time needed to complete the initial work. If the years seem too long, remember that he was dealing with thousands of drill holes made by hand. And he was dealing with uncertain appropriations from the federal government.

Help for the General's projects was on the way beginning about 1860 when Alfred Nobel began his experiments in applying the energy of nitroglycerine for explosive purposes. Engaged in developing a powerful blasting material, Nobel knew that all explosives are chemical reactions that break down compounds into highly compressed gases, very hot gases, and those gases expand rapidly. Heat plus compressed energy destroys whatever is around it. By early summer 1862, Nobel had shown, on a very small scale, that nitroglycerine could be used as controllable explosive, but the problem of a fuse remained. He initially employed gunpowder to detonate the nitroglycerine.

Yet nitroglycerine's power—the speed, concussion, and heat of its blast—encouraged its use, if for no other reasons than it released seven to ten times more force than gunpowder. If you were to drop a weight onto gunpowder, you would need 16 inches for that weight to fall to explode the gunpowder. Nitroglycerine? One inch. That energy, contrasted with gunpowder, meant fewer workers were needed, leading to fewer deaths and delays. That energy meant a briefer time taken to do its work, and all of those factors meant lowered cost.

But the newness of the explosive led to the three notable disasters in 1871. • Accidental explosions of nitroglycerine—Boston, Friday, January 27, 1871. • Another explosion again at Titusville, Hoosac Tunnel, March 1871. Mowbray himself wrote: "The new magazine had hardly been completed, and stored with nitroglycerine, when, on Sunday morning, at half past six o'clock, March twelfth, 1871, the neighborhood was startled by another explosion of 1,600 pounds of nitroglycerine. The cause of this last explosion was the continuous overheating of the magazine" ("Tales of Destruction."). • Nitroglycerine explosion Monday, May 22, 1871, Titusville again, where "Mr. Charles C. Clark … was blown to atoms as well as his wagon and horse."

The work too in the science of batteries proved valuable. In the 1850s, Maillefert had used a short-lived galvanic battery whose history goes back to 1799. Needing for it to rest completely level, a difficulty in a rowboat on the raging East River, the battery also generated a low voltage. Once more, Newton seized on the newest tools, in this case the discovery of Gaston Plante, who, in 1859 invented the lead acid storage battery, an improved dryer cell battery and once which generated slightly more power than the galvanic.

Even with the breakthroughs, the most efficient method of the lowering of rocks and reefs like Way's had yet to be solved. Most likely for two months during the winter no work on the river could be attempted. Newton also knew that because of the short period of slack water (the calmest time between the onrush of tides) meant that he would have to work during all of daylight, not just that half hour when there was a no (or little) tide running.

Surely rowboats, as Maillefert had used, and Newton was temporarily using, would not suffice. Then too, lowering the level of rocks to the required 26 feet meant that the rocks must be carefully attacked underwater using divers who would also have to contend with the power and unpredictability of the currents. If no new inventions nor tools were forthcoming, if the technology of drilling on and under the water was not available from other sources, Newton must create his own.

To purchase most of what he needed for demolishing the rocks and reefs in the river, two budget appropriations were approved so that by May 1871, during the 1870–1871 fiscal year, he was able to put together the materials and work force he needed. The sizeable amount of support, $178,200, had arrived for the East River Hell Gate project on April 10, 1869. On July 11, 1870, another $250,000 was made available.

# Six

# Newton's Marvelous Machine

## *The Creation of Newton's Steam Drilling Scow*

With the July 11, 1871, monies, Newton thought he would be able to solve some pressing problems first. These included the creation of a vessel sturdy enough to withstand collisions in the treacherous waters, a vessel stable enough from which safe and efficient drilling might be accomplished. Then too, the safety of the divers must be considered while they tried to do their work in the treacherous and powerful currents of Hell Gate.

The vessel to solve that problem had begun to be invented and designed by General Newton in the early summer of 1869. What was being built from the marvelous amalgamation first from the brain of Newton and then later by adaptations by both Newton and his superintendent Sidney S. Shelburne would soon be called a "steam drilling scow." The massive ruggedness of this new combination of a boat and a submarine drilling platform would perform duties for the government for many years.

Newton knew the scow had to be sturdy. Since the possibility was high of other vessels colliding with the scow in the turbulence of Hell Gate, the overall design of the scow had to emphasize strength and durability. Newton's plan had the scow with a deck at its thickest measuring two and one-half feet thick with a protective overhang to extend the hull by four feet. The overhang was covered with iron plating two inches thick. Finished, the scow was 127 feet long by 58 feet wide and 9.6 feet deep. It weighed 750 gross tons. If the measurements seem very large, it must be remembered that this vessel would be the platform from which drilling in 8 MPH currents would take place, and from which divers would descend and rise.

The steam drilling scow's stability was mostly accomplished using the anchors and the chains Newton foresaw would be needed in January 1867. The mooring chains extended out into the river and were attached to the anchors to hold the scow in place at the stern, bow, and center. The boat could not move very much at all.

On the following page is a view of the U.S. steam drilling scow. The

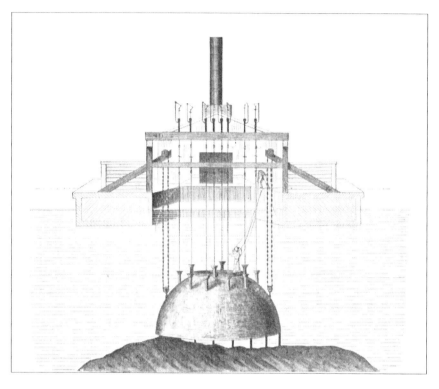

**The Newton Steam Drilling Scow in cross-section also shows the working of a number of drills at the same time, even as a diver shown here worked to position both the dome and its hammer drills.**

center of the scow was a well, built as an opening in the hull in the center of boat, of about 30 feet wide by 40 feet long.

This well would serve as the entranceway for the drills. Above the well, suspended by a derrick, a half dome of boilerplate made of iron on an iron frame. This was lowered into the well (and raised from it) by four chains, each chain connected to its own hoisting engine. The dome served as the framework for holding 21 drill-tubes, each with its own drill bit. The dome by its weight alone served as a fixed platform.

Into the tubes were placed drills with cross-shaped cutting edges. Weighing between 600 and 700 pounds each, they were able to create holes six to eight feet apart and not less than four feet deep. Through the tubes in the dome, the drills were lowered over "the point of attack. The engines raise these drills through a distance of 18 inches, and, releasing them, they fall upon the rock by their own gravity."[1] So these drills are percussion drills, sometimes referred to as "hammer drills." No steam drills were here, since "steam drilling" simply meant that steam power would raise the drill, and gravity, not rotation, would make the hole. Though they did not rotate, they had two advantages: first, a large area of rock might be attacked by 21 drills simultaneously; second, by careful plotting, any number in any configuration of drills might be put into service. Because of the tubes, the drills could be made stable. Another touch of genius from Newton.

After the scow's supervisor exactly plotted the position of the dome and the rock by using a sextant, the final steadying aspect of the steam drilling scow was described by its inventor as "a dozen stout legs, so arranged that they can be let go all at once, when one edge of the dome touches the reef to be operated on. The legs are held by self-acting cams, so that, when extended to fit the uneven surface of the reef the dome is to stand on, they are securely locked, and thus support the dome in an upright position."[2]

Thus, all the following features designed by Newton made for a portable drilling platform. • The legs of the dome are adjustable to the surface of the river's rocks. • The dome is heavy and circular allowing water to flow around it, not against it. • The heavy metal tubes encase the drills. • The chains hold the dome. • The massive scow it secured in place by chains and anchors.

Newton's "average working force of the United States steam-drilling-scow, while engaged on Way's Reef, numbered thirty-seven men, consisting, besides [Mr. Striedinger], of one mechanical draughtsman, two divers, one chief carpenter, two carpenters, one engineer, eight drillers, one blaster, one blacksmith and two blacksmith helpers, twelve sailors, two firemen, one time-keeper on the dredge, and one tide-gauge keeper on board of the scow."[3]

The construction of this scow-drill machine was commenced in July 1869 and finished in 1870. The practical work was begun in May 1871, upon Diamond Reef, near the mouth of the East River. Now Newton had to make certain to use the scow to maximum benefit and safety.

Newton's steam drilling scow was still in use after his retirement almost 20 years later. At the World's Columbian Exposition in Chicago in 1893, the War Department Exhibited Newton's Scow, writing

> By the above described machine, Ways Reef, Shell Drake, and North Brothers Island Reef, in Hell Gate; and Diamond Reef, Coenties Reef, reef off Diamond Reef, and Pilgrim Rock, in East River, have been removed. It has also operated upon Frying Pan, Pot Rock, and Heel Tap Rock, in Hell Gate; and Ferry Reef, in the East River; Baxters Ledge, in New York Harbor; Corning Rock, in New Rochelle Harbor, and upon a channel through a reef in the Harlem River.[4]

A report describing Newton's Scow was entered into the records of the House of Commons in England.

These projects lay some years ahead. At the same time as the scow,

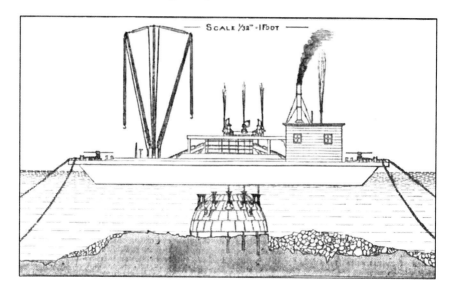

SCALE ³⁄₃₂" -1 FOOT

An artist's version of the Newton Scow provides us with a scale to get some idea of the size of the vessels and its workings. The ropes (actually chains) extending from the stern and bow were secured to hold the vessel in place while it worked within dangerous currents, strong tides, and the presence of other vessels.

with a Mr. Pearce appointed supervisor, was being built, the work at Hallet's Point had to be accomplished as well. For now, how specifically would Newton deal with the methods and techniques for handling and transporting the nitroglycerine, the dredging and dispersal of the rock debris, and the employment of divers and diving equipment.

Newton had to find a way to use his appropriations to buy four items. To handle the dredging, he must purchase two floating derricks to haul up from the river floor the exploded rock, and to haul it away, a scow with a 50 foot mast and hoisting gear.

Many more scows were needed and 24 of them of varied sizes were purchased. Three stone-scows, each of about 250 tons capacity, were put to work in this fashion: while one was loading, the second was unloading, and the third was on its way to or from. The stone-scows were always measured empty and filled, to compute, by means of their displacement, their actual stone-load.

To run errands, to transport materials, and to move workmen from the shore to Newton's Scow, six small boats became part of Newton's small fleet. In addition, the General included the need for buoys to warn off ships in the area from his scows, and money to pay for repairs on machines, to sharpen drill bits, and to replace anchors and chains due to collisions.

## Nitroglycerine Is Studied

But it was still true that no matter what equipment was readied, what was most necessary to the work on the obstacles did not yet exist. Work on the Frying Pan could proceed with hand drills and nitroglycerine. But if hand drills and nitroglycerine were used at Hallet's Point tunnels the work would take a very long time and continue to be dangerous due to the volatility of nitroglycerine.

Around the same time, General Newton could read in both popular and scientific journals such as *Journal of the Association of Engineering Societies, The Royal Society Proceedings* from Britain, and *Chemical News* about various experiments with nitroglycerine, and keep track of advances in handling the volatile liquid. Then too the work of General Henry L. Abbot at Willet's Point school was adding practical information about explosives useful in demolition and defense-related projects. Using September 1870 tests of blasting experiments performed in the river itself, Abbot was able to pass data on to his colleague Newton.

What needed to be learned about nitroglycerine? For budgeting purposes first, what amount of chemicals needed to be bought? How would that be determined? Experimentation and testing would demonstrate the effectiveness, the power, of nitroglycerine particularly as contrasted with gunpowder, the explosive of Maillefert. How will the explosive be detonated?

The power of an explosive can be measured in many ways. To begin with, the explosion, in this case, to shatter rock, must produce sufficient pressure sometimes measured in pounds per square inch (PSI) to do the job required of it. Air pressure is measured in "atmospheres," normally 14.7 PSI. Gunpowder produces 6,000 atmospheres or 76,241 PSI, but no shock wave. Nitroglycerin developed a pressure of 20,000 atmospheres, 294,000 PSI or 1,360 times more than the air pressure in Newton's office at No. 7 Bowling Green. This shows that nitroglycerine makes almost four times more (3.85) than the pressure than gunpowder does.

Another quality of the explosion is heat, a trait when transferred to colder objects, like from a cutting torch to a piece of steel, breaks or splits the colder object. The heat given out by the combustion of gunpowder is 1,145° F. For nitroglycerine the heat liberated raises the temperature to about 9,000 degrees, 7.8 times greater than gunpowder.

Next is speed. In a falling object demonstration taken from the television show *Mythbusters*, a piano weighing 700 pounds and dropped 50 feet before it crashed would be traveling at 38 MPH and have an "impact force" of 12,000 pounds. The change from 0 MPH to 38 MPH increases the effect of the crash on the piano. Gunpowder's burn rate has a speed range, on average, of 894 MPH. Nitroglycerine's detonation wave moves at 4.78 miles per second or 17,208 miles per hour, 19 times faster than gunpowder.

So, were we to imagine a giant sledgehammer (the pressure) moving, instead of its about 11 miles per hour, moving at 17,224 miles per hour and producing 9,000 degrees of temperature, we have a small idea of the tremendous shattering force exerted on rock by nitroglycerine.

Though dynamite had been patented by Nobel in the United States since May 26, 1868, Newton knew that "[f]or submarine work nitroglycerine is unquestionably the best explosive known." One obvious reason is that since it is a liquid, it cannot be tamed by water as was the earlier gunpowder. Here it can be seen that rather than placing kegs of gunpowder, holes can be drilled in rock and filled with the explosive nitroglycerine. Since nitroglycerine is used, smaller holes than those for gunpowder were needed, cutting back the drilling time for the holes.

Newton and his staff, along with the experimenters at Willets Points, were always searching for a better (and cheaper) way. For a few years, gunpowder was used to detonate nitroglycerine. But then Nobel in 1867 had patented the use of mercury fulminate as igniting caps for detonating the nitroglycerine. The blasting cap set off by electricity from the new batteries would provide an immediate jolt to the nitroglycerine or any product that included nitroglycerine. Now the charges could be fired with effectiveness and reliability.

## The Handling, Transportation and Storage of Nitroglycerine

At some point early in his work, Newton realized the virtue of having a large rock in the river to serve many functions—chemistry lab, storage facility for tools, etc. This was Great Mill Rock which stood next to Little Mill Rock, 1,600 feet off the shoreline at Manhattan's 94th Street. Covering more than eight acres, Mill Rock showed the desirable qualities of isolation and size. The general also used the space to conduct more tests of drilling and blasting, tests that made Newton understand that newer methods using better drills and more powerful explosives would lead to much lower costs than he had calculated originally in 1867. In one instance, the cost had dropped from $65 per cubic yard to $15. Newton, demanding and careful—he was working in between two large sections of New York—hired two men, Mr. William Pease, assisted by Mr. F. Sylvester, to conduct detailed surveys of the work to be done. They took more than 16,000 soundings, each separately located from the shore, by means of instruments. It often appears in various newspapers that Newton is labeled as the "able and present" topographical engineer, one whose military work encompasses "terrain analyses, terrain visualization, nonstandard map products, and baseline survey data."

Those experiments on Mill Rock and by the Corps of Engineers at Willets Point with the new demolition chemical were surely needed. Sometimes called "chemical oil," nitroglycerine established its presence in the public mind though accounts of seemingly random detonations. One reason for the volatility concerned transporting nitroglycerine. With jostling, small bubbles are created, which are then packed together when the substance is simply in motion. This compression causes a detonation even when the motion is slight.

Newton and his superintendents and assistants continued to learn the best ways to buy, transport, and handle nitroglycerine. Newton had written that the danger of nitroglycerine can be minimized if "it is carefully compounded with pure materials, kept free from conditions unfavorable to its stability, and used immediately."[5] This *Scribner's Monthly* article continues in saying that the nitroglycerine used in the beginning attempts on East River rocks was carefully handled, the operation and equipment "used was designed by Mr. Pearce, assisted by Mr. Paul Marcelin, who, in connection with Mr. Warren, conducts the manufacture."[6] Newton, forced to exercise great caution as the work went on in the crowded city, chose a nitroglycerine factory not on shore, but in the shallows, four feet of water, a mile or two south of Communipaw (part of Jersey City, New Jersey). "Since the factory was in the water, boats can be moved directly from the factory to the points where it had to be used. The boat containing the explosive arrives from the nitroglycerine factory, towing a small boat which flies the red flag of danger. In addition, the explosive is never kept more than a few days, and always at a temperature slightly above freezing-point."[7] But this situation was still early in what could yet be learned.

Even as Newton's work progressed, Hell Gate remained treacherous. On July 8, 1871, the schooner *M.A. Longbery* bound from Elizabethport, New Jersey, to Bridgeport, Connecticut, sank at Hell Gate. Three weeks later, the sloop *Thomas Ransen* bound to New Haven with a cargo of coal went ashore on Holmes Rock (off the south end of Randall's Island), and less than a month later, August 21, 1871, the schooner *Juno* of Rockland, Maine, with a cargo of lime, ran onto The Gridiron (close to Astoria), Hell Gate, caught fire and was a total loss. In September, in the space of 12 days, three more ships were lost.

## Newton Adapts to the Steam Drill

Practical calculations could now move forward. Newton also decided to buy another technological breakthrough, beginning in May of 1871, putting the new steam drills to work on Diamond Reef. Patented that same year, Simon Ingersoll's invention marked a breakthrough in that the drill's pistons were cooled by water. Before Ingersoll, overheated pistons meant the drill would have to take time to cool before taking up work again. Some of that reef's 800 cubic yards of rock were soon turned into debris, which needed to be grappled and taken away on the scows.

Perhaps it was thought that the new tools coming along might speed the work that had been started at Hallet's Point. Newton could not know it at the time but five more years would pass before Hallet's Point could be destroyed.

At least by November 1, 1871, Hallet's shaft drilling was completed with limitations and fortunately so. All mid-river work stopped for some months because the East River froze, stopping traffic on the waterway, halting commerce via that river, and delaying the work of Newton and his scow. But the completion of the Hallet's shaft meant that the work of tunneling under the river could continue through the winter. With Hallet's Point Reef connected to Astoria, there were no delays in bringing in material and workers.

For the most part, the rock, foliated hornblende gneiss with numerous veins of quartz, afforded little progress though work continued on Diamond Reef and Coenties Reef (at South Street), both well south of Hell Gate. There, new, more exact knowledge and more efficient apparatus were beginning to be made use of in Newton's tasks, particularly for blasting and drilling.

Nevertheless, at around the same time, funding slowdowns—the appropriation of 1871 totaling $225,000 equaled just one half the amount asked for by the General—caused Newton to suspend operations for months on the East River, most especially on The Gridiron section of Flood Rock. Newton had other supervisory work, funded differently, required of him from the Delaware Breakwater Harbor of Refuge to a survey of Otter Creek, Vermont, about 450 miles away.

## *The Railroads Show Their Promise*

While Newton's work continued as best it could for river traffic, at the same time the trains kept running into Manhattan. And that same year, 1871, Vanderbilt's Grand Central Depot opened at Fourth (now Park) Avenue and 42nd Street, serving as the terminus for the New York Central and the New York, New Haven & Hartford, which served southern New England. The iron and glass train shed rose 100 feet and enclosed 15 tracks. Then one of the largest interior spaces in America; later it would be made even larger.

The majority of railroads were dependent on the waterways to take their passengers and freight into Manhattan and Brooklyn. This artwork shows how the Pennsylvania Railroad needed to have a Ferry Station at Courtland Street from at least 1871 until 1910, when the tunnel from New Jersey was finished and Pennsylvania Station at Seventh Avenue & 32nd Street opened (New-York Historical Society).

Other vast railroads, the Erie and the Pennsylvania, for example, had their terminals on the other side of the Hudson in New Jersey. The terminals there were really just waiting rooms and train sheds, because once more, it was water that delivered people and goods in the first and final leg of the journey. Passengers and freight must ferry their persons and ship their cargoes across to the New York Port.

The Pennsylvania Railroad had to build the ferry station in New Jersey and on the other side of the Hudson. It built slips in New York City to receive passengers and provide them with a covered space when departing.

But no matter how fast the railroads extended out into the country, it was commerce by water that impressed. In 1871 customs officers just at the port of New York collected $141.4 million, when total interest obligations of the federal government added up to $125.5 million dollars.

## Hallet's Point Reef: The Tools and the Workers

Meanwhile, at Hallet's Point, more newspaper coverage began to appear. The newness of the nature of attempt Newton was making attracted the press even as the size of the project continued to grow. The work at Hallet's Point was identified as "tunnel blasting," as done "by the efforts of 82

Cornish miners" who "dug a gigantic hole 28 feet deep and 300 feet in circumference excavated through the solid rock."[8] The amount of rock excavated from these horizontal tasks that year came to 8,306 cubic yards, and the drilling was mostly done by hand.

True, much of this work has been accomplished by these miners from Cornwall, whose experience went back centuries. Mining in tunnels under the Irish Sea made possible work to be done at 700 yards below sea level. Using gunpowder to blast the rock, the first steam-powered rock-boring machine and a timed safety fuse were among the advances at Cornwall mining. But in the mid–1800s when the lowered price of both copper and tin made its mining in Cornwall unprofitable, many Cornish men and women, seeking work, immigrated to Canada and the United States.

The Ingersoll percussion-machine, a rotating tool, used the popular stable tripod design, whose patent was granted on March 7, 1871. Rudolph Leschot was permitted a patent in 1863 but it was not until 1869 that a working diamond drill bit was shipped to the United States. The third of the drills, which all seemed to appear at the same time, was the Rand drill, a steam-powered version. For this drill to work, needed was a steam engine and a boiler, already long available.

Once more General Newton was willing to try any new tool that might move work more expeditiously. Using the best of these tool advances, it is known that by November 1, 1871, Newton could report "completion by hired labor of shaft at Hallet's Point and 709 linear feet of tunnels, with 182 linear feet of galleries removing 8,293 cubic yards of rock [22,391,100 pounds]."[9] These numbers account for about 12 percent of the complete work.

That "hole" that was the shaft had reached a width of 50 feet. At the shaft's bottom, behind the coffer dam, a space was made for workshops of iron and wood, to serve as places for carpenters, engineers and stone masons to prepare their work. At the shaft's open mouth sat a derrick 60 feet high which was used to haul up both rock and workmen. In place were five steam engines.

New Yorkers curious about the work were pictured strolling around the coffer dam and looking down into the shaft, women with umbrellas held high, in a *Scientific American* issue late in 1871. What the visitors could not see were the galleries, horizontal lanes that intersected with the vertical tunnels, tunnels that would soon extend far under the East River itself.

## Interested Parties

What this year proved, not so much differently from all the other years, was that with so many work-intensive jobs to do and with such limited funds, and with new technology continually appearing, Newton, in the newspapers, seemed to be moving from one job to another. The Newton Scow itself would appear, for example, at Overslaugh Rock (about 80 percent of the way up the Hudson) or down on the East River in no discernible pattern.

For General Newton, there continued to be multiple demands from multiple interest groups and agencies and from their influential leaders. These groups can be listed:

> House of Representatives Committee on Commerce (1819–1892)
> The Commerce and Manufactures Committee (1825–1946)
> The Committee on Roads and Canals (1831–69) and its successor the
>     Committee on Railways and Canals (1869–1927)
> The U.S. Army Corps of Engineers
> The House Appropriations Committee
> The War Department
> The New York Ship Owners Association
> The Committee on Naval Affairs
> East River Merchants Association
> The New York City Department of Docks
> The Commissioners of the Dock Department
> Supporters of the Harlem River Canal
> The Chamber of Commerce of the City of New York

In late September 1871, the steam drilling scow moved from Pier 8 on the East River (around 55 Water Street) up to just south of Randall's Island during work on Coenties Reef, Shelldrake, Way's, Hog's Back, and Pot Rock. But some parties that called on Newton wanted only to inspect the works at Hallet's.

Boarding the steamer *Henry Smith* at the foot of East 23rd Street was an assemblage consisting of the Grand Duke Alexis of Russia on a tour of the United States, the Secretary of War, the commanding general of the Army William Tecumseh Sherman, General Humphreys, Chief of Engineers, General Abbot, from Willets Point (an aide-de-camp to Humphreys during the Civil War), and others along with General Newton.

At Willet's the party saw demonstrations of various underwater explosives at Hallet's, the group descended the shaft. Hallet's was the Duke's second stop during his visit to New York City, probably on November 21, 1871.

## *The Danger Remains; the Funds Arrive*

Looking back a few years from 1872, what was true then was true in 1872:

**So famous was Newton's work on the East River that visiting dignitaries asked to be taken to the site. Pictured here is the Grand Duke Alexis of Russia viewing Newton's work in 1871 (*Frank Leslie's Illustrated Newspaper*, October 7, 1876).**

By the 1850s, one in fifty ships passing through the Hell Gate were either damaged or sunk—an annual average of 1,000 ships ran aground in the strait. Ships would navigate extra ocean mileage to avoid this passage to the Atlantic. Captains seeking to test their mettle would have to wait for the slim window of time where they could navigate their ships safely through the Hell Gate.[10]

But once more the problems remained of getting cargo into Manhattan by approaching the New York Harbor from the Atlantic Ocean at Sandy Hook. The treachery of the passage and the concomitant expense caused many owners and captains to choose the entrance via Long Island Sound and the East River, even with Hell Gate's threats. In addition, though not part of Newton's portfolio,

New York continued to grow as the financial and commercial center of the United States, and more goods were delivered to the city. But ... because of varying depths of water in the harbor and a lack of standardized dock sizes. Ships often weren't able to get in close to the shore and may have had to offload their cargo onto smaller boats, or may not have been able to trade in the city at all."[11]

In April of 1872, the River and Harbor Improvement bill passed in the House of Representatives, appropriating about $225,000 for Hell Gate. That spring, the work on all of the smaller obstructions went on, with much of the work under the immediate management of sub-contractors signed with the Corps of Engineers. As the years went by, separate arrangements would have to be made with the drillers and the workers who removed the debris.

The Newton steam drilling scow, towed by a steam tugboat, might be taken all the way down the East River, almost to Governor's Island, to work on Diamond Reef and then moved all the way up past Hallet's Point to Pot Cove in order to work on other obstacles.

# Seven

# The Scow at Work

## Progress with Some Obstacles in the River

Most of these lesser obstacles were worked on using the steam drilling scow at various times for more than a decade. Other obstructions had been worked on, some briefly by Maillefert. Two and a half decades earlier it was true that Maillefert had blown Bald-Headed Billy and Hoyt's Rocks into deep water. Some rocks and reefs still exist. For example, marked on a Google map in 2020 north of Hell Gate one can see Rhinelander's Reef, Mill Rock, Hell Tap, Hog Back, Holmes Reef and Way's Reef. They all are outside of the normal passage of vessels through the area. (A complete list can be found in the appendix to this book.)

One of the smaller obstructions included in the early work on the river was the Frying Pan, measured in 1867 at 12,475 square feet. Found halfway between Hallet's Point and Ward's Island, the rock's work estimate reached 51,675 cubic feet of rock to be demolished, a size equal to 139,522,500 pounds of rock. The other obstacle was Pot Rock between Ward's Island and North Astoria, just west of Pot Cove. The debris from that blockage totaled 90,450,000 pounds of rock to be grappled and taken away. (Each cubic yard of rock weighed about 2,700 pounds.)

Recognized as a fierce place, Pot Rock's problems were accurately described in 1852 in the periodical *The Merchants' Magazine and Commercial Review*, which wrote, "the violent agitation of the water above and around Pot Rock, and the wild roar which accompanied it, was exactly such as if some sea monster were struggling in agony, vainly attempting to reach the surface of the water."[1] When the tide was running, Pot Rock could not even be approached in a small boat, and the only available time for blasting it, was during slack water. But slack water lasted only some ten minutes—never more than ten—and the operations were confined to that limited space of time.

Newton later remembered "work was prosecuted on Pot Rock in Hell Gate from August 5 till December 28, 1872, during which period the scow was much exposed to collisions, of which sixteen took place. In one of them the colliding vessel was drawn under the scow and carried off the dome, which was afterward recovered, considerably damaged, in eighty feet of water."[2] But soon the depth on this rock was now 24 feet.

Way's Reef was considered to be sufficiently lowered later (in fiscal year 1874–75), with Coenties Reef lowered in the period 1875–76. Still with the aim of decreasing the hazard of these two exposed rocks until a depth of 26 feet was reached, work continued through thousands of cubic yards of rock being drilled, blasted, dredged and taken away.

The larger Hallet's Point Reef had been a serious obstruction in the East Channel, dangerous to large and small craft alike; it did not leave enough sea-way for vessels floating down with the ebb and steering clear of Flood Rock, and ships had often been smashed against it. It also created dangerous eddies at either tide and a strong drift toward the Frying Pan.

## Changes in Ships Verify Newton's Work on the River

Even as the work on lowering the rocks went on, shipping changes became more apparent, changes that Newton needed to take into account. By 1870 the only shipbuilders left in New York City and vicinity were Lawrence & Foulks, Thomas Stack, Webb & Bell, and John English & Sons. Even the famous William H. Webb yard was idle, and others went into different businesses like paint. Webb had produced 136 ships of all types, launching his last ship on May 26, 1869.

The United States had been left behind in steam vessel production. A lack of foresight, the U.S. Civil War, and the excellence of their clippers all contributed to the decline. During the Civil War, more than 100,000 tons would be sunk by Confederate raiders, while other ships changed their American flags to those of other countries to protect them on the high seas. Though before the war the U.S. could tally almost two and a half million tons of cargo transported, while the Civil War years saw a loss of 860,000 tons. By 1879 it was well-known that of "about sixty-five percent of America's imports and exports, about half had been carried in American bottoms. War cut this percentage almost in half. America had lost control of the carrying trade for even her own commerce."[3] By 1880, the port of New York would see 2,180 American sailing ships. Of steamships, 1,501 were foreign while only 186 registered as American.

In 1872 the federal government chose to subsidize the building of iron ships and the winner of the contract was John Roach and Sons in Chester, Pennsylvania. After upgrading a fleet of side-wheelers, they built two iron sister ships of 4,000 ton, one named *City of Peking* and the other *City of Tokio*. (For contrast William Webb built a "medium clipper" of 1,750 tons in 1850.) Iron the ship might be, but it still had masts.

Around this time Boss Tweed of the city's Tammany Hall was arrested and the power of his organization declined. Tweed himself went on the run for four years. General Newton, a federal employee, was able to deal with companies in New York City without having to depend on Tweed or Tammany. Newton's reputation remained as an efficient and honest engineer who reported on every inch of wire, every pound of explosive, every salary of every employee of the scow, both for the work with the scow on the river and at the Hallet's Reef project.

## The Importance of Julius H. Striedinger

It must be clearly pointed out that the work on Hallet's point as well as the work on the river would likely have been of a much different kind without Newton's hiring of Julius Hermann Striedinger. Then in his early thirties, Striedinger went on to publish at least ten scientific articles having to do with electricity and detonation, such as "Igniting Blasts by Means of Electricity" and "Blasting with Nitroglycerine." Clearly, his most important work—both to science in general and to Newton's work on the East River—was his 1877 article "On the Simultaneous Ignition of Thousands of Mines and the Most Advantageous Grouping of Fuses." One year later Mr. Striedinger, at $2,000 per year, was appointed to be "Instructor of the Corps of Sappers and Miners" for the New York Fire Department in the use of chemical agents to blow up buildings to prevent the spread of large fires. Newton himself, in 1875, wrote to "express my sense of the skill of Mr. Striedinger displayed not only in the management of the scow and drilling apparatus and likewise in the manipulation of the explosive agent used."[4] Newton's great assistant would serve from 1873 to 1875.

There remained much to learn. There were still improvements being tried for diving and for the safety of the divers. The regular operating procedure for the placement of the explosives in the drilled holes in the rock had not yet been made standard. Work continued even as experimentation went on, still looking for the safest and most efficient ways of handling and placing the nitroglycerine.

## Newton's Refining of the Scow's Operations

Newton's working scow had been struck by the steamer *Providence* on August 3, 1871, and again by the same ship on October 15, 1872. These collisions caused the authorities to force Newton to "suspend work once more because it was clear that the 1860 regulations were not being observed by ships' pilots." These pilots, by state law, were charged with "the safe pilotage of vessels through the channel of the East river, commonly called Hell Gate."[5] Due to the size of the Newton steam drilling scow and how it is anchored, the pilots thought of this working vessel as an impediment to their trade. Some of the pilots thought the problem lay with the Scow in effect attacking their vessels.

Even as the work continued, the problems of ships trying to gain New York Harbor from the Atlantic Ocean would continue to cost money. It was the job of the Sandy Hook pilots to take vessels from Sandy Hook to the point of the current Throggs Neck Bridge, and the job of the docking pilots to take ships from the Bridge to their berths in Manhattan.

Because nitroglycerine was now being used all around Hell Gate, Newton's explosions became louder and the giant-sized scow became a normal presence on the river, and so the newspapers sent writers to explain and entertain their readers. Accidents had happened in many places already throughout the country and world, and the prevailing thought of editors was that more accidents were bound to happen.

Some newspaper reporters chose to describe the scow's tasks—for example positioning the craft for work. Using a rope line attached to the center of the dome, a weight was lowered onto the target rock. Then a tug pushed the scow until the weight and line marked the center of the well on the scow. The scow was then in place to begin drilling and only then was the dome lowered for drilling. "At each blast, the rock is broken up to the depth required, over an area of four or five hundred square feet" an 1872 magazine reported. "When the whole reef has been gone over so that there is no place for setting the dome, the scow is hauled off and the broken rock removed by a huge grappling machine."[6]

When these maneuvers were used on Pot Rock, they lowered its depth to 24 feet. With the blasting completed, the power of the nearby whirlpool diminished, and words of admiration were heard for Newton, citing his persistence and ingenuity.

While work with the scow and work on the shaft, tunnels and galleries at Hallet's Point had been going on without incident, in Michigan, Pennsylvania, New Jersey, and elsewhere in New York, injury, destruction and death had occurred, and not just in 1872. The first headline read "Nitroglycerin Explosion, Seven Men Seriously Injured," July 10, 1872, in Kearney Township, Michigan. Two workers listed in critical condition.

The *Titusville (Pennsylvania) Morning Herald* headlined on August 27, 1872, "Blown to Atoms…. Another 'terrible nitroglycerine explosion took place near Rouseville yesterday morning at six minutes after eleven o'clock, by which Mr. Wm. H. Payne, torpedo agent at that place, was instantly killed and blown into diminutive fragments.'"

"A Nitroglycerine Explosion, Several Men Injured," October 6, 1872, near Bayonne, New Jersey. While men were experimenting with nitroglycerine, one explosion sent "granite boulders flying through the air, one over a thousand pounds, fell through the roof of a house 200 feet distant. Two of the men were fearfully mangled."

Then "A Terrible Accident." November 26, 1872, near Yonkers, New York. "Two Young men Blown to Atoms, and Two Others Seriously Injured, by Nitroglycerine Explosion" when it was discovered that all it took was young men throwing stones at stored cans of nitroglycerine. "Intense excitement now prevails in the vicinity" once it was known how casually the nitroglycerine was being handled.

## The Accidents Continue and the Tunnels Move Under the River

By Newton's sixth year on the Hell Gate project at Astoria and elsewhere, he can point again to the river's appetite for the destruction of valuable ships and their valuable cargo. From May 10, 1872, ships like *The Morning Star*—and six others—foundered on the rocks and reefs of Hell Gate. On May 2, the schooner *William R. Knapp* was rammed by the steamer *City of Hartford* between Hell Gate and Astoria. The whole of her stern was carried away and she sank. On 10 May the schooner *William Butman*, bound for Boston from Elizabethport, struck a reef and sank. On July 20, 1872, the schooner *Diadem*, loaded with coal, sunk in five minutes after colliding with the steamer *Galeta* off Ward's Island. Soon, the wreck of the *Diadem* was struck by a schooner named *Flagg*. It capsized; the crew were saved. Others crashed into rocks but did not sink. They were towed away and put into dry dock. On August 18 the schooner *Black Diamond* with a cargo of coal struck on North Brother Island, Hell Gate, and sank. On August 28 the schooner *C.L. Hulse* with a cargo of Delaware & Hudson coal, collided and sank at Hell Gate. In August, four more ships were destroyed.

Even paddlewheels steamers were not safe from the vagaries of the River. The steamer *City of Lawrence*, the first (1867) Long Island Sound paddle steamer with an iron hull, weighing 1,351 tons, collided in the East River on September 3 with the schooner *Empire State*. Newton too was affected by the unpredictability of Hell Gate and during the Scow's labors at Pot Rock, 16 collisions occurred; two of the vessels were sunk, and one of them was driven under the scow.

Meanwhile, in New York Harbor off Sandy Hook, ships were also victims of the usual problems of running aground or sinking. So Hell Gate still made the most sense.

General Newton's responsibilities had become more far ranging and as assistants he took on two men, one identified in reports as F. Sylvester, the surveyor. Meanwhile, the Corps of Engineers took on more projects, all over the country, while at nearby Hell Gate, Newton oversaw the building of a lighthouse on the northern end of Blackwell's Island, ordered a survey of the Upper Harlem River, and a survey of a transportation route from Troy to New York City.

With the vertical shaft at Hallet's Point now completed, the drilling, blasting, and removal of rock could continue to move outward—that is, under the East River. Each tunnel was designed early in 1870 and Newton's plan was to extend the tunnels between 150 and 350 feet under the Point, with cross galleries every 25–30 feet. End to end, the most distant tunnels were separated by 880 feet. Tunnels measured 10 to 15 feet wide and 20 feet high and were named Farragut, Humphreys, Madison, Hoffman, Sherman, Jefferson, Grant, McClellan, Franklin and Jackson.

## Improvements in Steam Drills and Other Tools

Blasting in the tunnels could not be a rapid process since charges of fewer than eight ounces of nitroglycerine were generally selected. In the early days of Hallet's tunneling, the Bickford Safety Fuse from 1831 was used to set the explosives off, a fuse that required putting a flame to the explosion.

It has been established that work on fuses and cartridges were often left in the hands of civil engineer Julius Striedinger. Striedinger, through experimentation and testing, chose always to use nitroglycerine or compounds of nitroglycerine as the explosive material. In spots where the rock "presented as thin sheets ... gunpowder would suffice." His best choice for the cartridges placed in the drilled holes was to construct them of pasteboard, coated with "an impervious composition," and the percussion supplied either by gun cotton or by a cap containing fulminate of mercury. But beyond those duties Newton wrote, "I am indebted for the careful compilation of the facts in regard to the tides and currents, and to the obstructions requiring removal."[7]

The site by now employed 100 workers working two shifts per day—8–4 and 4–midnight—each paid $2 per day. The costs for each of the operations to complete the task were calculated by percentages in this way:

| Blasting 46%                      | Pumping 10.37%      |
|-----------------------------------|---------------------|
| Conveying rock to shaft 17%       | Incidentals 21.32%  |
| Hoisting 13.28%                   | *Total* 100%        |
| Dumping 2.03%                     |                     |

But when there was no work on Sunday or in good weather the site was visited by an average of 3,000 people; on weekdays 200, who knew they needed to bring some device for lighting as well as an umbrella. If they ventured down steps into the work area behind the dam or into the tunnels they would need rubberized coats and boots. Boats from other parts of Astoria and horse cars from Williamsburg took visitors there.

At the Hallet's Point site, to complete the tunneling, estimates said 50,000 to 60,000 cubic yards will have to be removed even with 9,554 cubic yards already mined. Once more Newton seized upon a newer tool for the work. After that breakthrough in drilling technology by Rodolphe Leschot with his diamond core drill bit, it seems clear that by 1872 chiefly used were six Burleigh drills (patented 1866) which punch and twist at the rock. This new Burleigh drill's chief advancement lay in that no longer would a drill simply hammer at the rock. Now Newton could report the increased speed of drilling but also had to admit that the drills were not known for their durability. The drill bit speed estimate at the time was calculated at 300 to 500 revolutions per minute, advancing perhaps one and a half inches. It can be supposed that when the drills broke down, then workers would go back to drilling by sledgehammer.

General Newton mentioned the Diamond, Hand, Winchester, Wood, and Waring drills, also worked by compressed air, which were still being tried out at Hallet's Point.

Just uptown from Newton's headquarters was the office of the Ingersoll Rock Drill Co. at 1½ Park Place. Once Simon Ingersoll invented the pneumatic drill in 1871, three distinct improvements were added to the drilling process. The drill itself rotated the drill bit so that rather than simply pounding at the rock, the drill would dig into the rock creating smaller debris rather than larger material. Second, Ingersoll followed his invention the next year with the improvement of cooling the cylinders with water, meaning the drilling could continue uninterrupted without waiting for the bit to cool and stay undamaged. Lastly, the drill was mounted on a stable tripod, at times called a drill carriage and made by a company in Birdsboro, Pennsylvania. In the case of Hallet's Point project, the tripod was then mounted on wheels and set on tracks. Once the drill was thus stabilized even further, it could be used on the ceiling of the tunnel.

2

**Now with steps in place at Hallet's, clearly to be seen are the East River, the Coffer Dam, tunnels, and rail tracks. One tunnel is blocked off, suggesting the placements of explosive cartridges there was accomplished.**

Lastly, once compressed air replaced steam, "this compressed air serves the purpose of a motive power equally as well as steam; and more than that, it not only drives the drill but cools the atmosphere, thus giving the workmen substantial comfort and fresh air."[8] In addition to safety, the history of the Hallet's Point mine illustrates the comparative costs of hand and machine drilling. Drilling by hand, by sledgehammer, cost 95 cents per foot; drilling with compressed air was 37 cents per foot, a 39 percent drop in cost.

The demand for pneumatic drills was driven especially by miners and tunnelers because steam engines needed fires to operate and the ventilation in mines and tunnels was inadequate to vent the fires' fumes; there was also no way to convey steam over long distances (e.g., from the surface to the bottom of a mine); and mines and tunnels occasionally contained flammable explosive gases such as methane. By contrast, compressed air could be conveyed over long distances without loss of its energy, and after the compressed air had been used to power equipment, it could ventilate a mine or tunnel.

Added were some pieces of equipment to make the drilling and rock removal more efficient.

Car-tracks in the tunnels and shaft ... have been laid. The broken rock was then loaded upon small cars ... which were hauled by mules. A small dummy-engine [at the top of the shaft], for hauling the excavated rock to the dump-pile was constructed. There a derrick worked by steam hoisted the car body and its contents from the track to the top of the shaft, and the rock there dumped into a tipping car attached to a small steam locomotive, which conveyed the debris, and deposited them at the end of the dump pile.[9]

This small railroad, never pictured, carried away the mostly pulverized rock—that is what the nitro does to rock—since the pieces of debris are too small to be used as filler somewhere else. Eventually, 2,000,000 cubic yards had to be blasted and then removed from the river bottom, a weight of six million pounds.

With improvements in drilling now available, tunnels progressed at 250 linear feet per month. The tunnels were kept free from water by constant pumping; in some of these the floor was inclined to drain off the water, pumped out at a usual rate of 1,000 gallons per minute, though pumps could handle 18,000 gallons per minute if needed. Tunnels were smoky from the workers' lamps. Once the tunnels had moved some distance under the river, Newton left in place what came to be call "piers." These are columns that support the roof of the galleries and tunnels because in some spots only ten feet separate the roof of the tunnel from the floor of the river. Steamers overhead could not only be heard but felt. Caution was enforced in the pounding from the drills and in the use of explosives to not disrupt the balance between river bottom and roof.

The greatest rate of progress under the river occurred during fiscal 1872–73, during which time 2,731 feet of galleries were driven, and 9,554 cubic yards of rock (25,795,800 pounds) were removed.

## EIGHT

# The Long Depression Begins

*The Work Goes On Even
as the Appropriations Fall Short*

For Newton's use in 1873–74, Congress appropriated $225,000 for Hell Gate, and the chief engineer reckoned the rock identified as The Gridiron would have its danger lessened by the installation of "spring fenders." The Gridiron remains to this day a mysterious obstacle. Seemingly included as some sort of notable part of Flood Rock on a large map drawn by Newton, any mention of The Gridiron disappears from the Annual Reports after 1872. This suggests that a fender was used. Yet since the passageway where Flood Rock once lay would be cleared by 1885, then the Gridiron was gone as well. It is lost in the 150 years since Newton's early work just as the location of Scaly Rock and Battery Reef have been lost, and such is the case with the location known as Sunken Meadow, a spot once separate from Randall's Island but now part of it, the space between filled in.

On May 13, 1873, the *Hope*, a steamer, bound on a short run from Blackwell's Island to Hart's Island, was run down at Hell Gate by the steamer *Americus*. The *Hope* was cut in two and four men were killed. The wreckage was not removed for years and caused many navigation problems on the river. On January 25 the schooner *Charles A. Grainer*, Port Johnston, New Jersey, for Providence, sank at Hell Gate; on September 7 the steam tug *Vixen* was run into and cut in half by the steamship *Granite State* near Hell Gate.

Meanwhile, steamers traveled on the East River to transport passengers to and from smaller ports while using Long Island Sound most directly. Docking was made available at Oyster Bay, Stamford, and Hartford among about seven others. Other ships looking to avoid delays at Sandy Hook or trying to avoid the cost of lighters to unload chose the perils of Hell Gate to arrive from major ports such as Havana, Galveston, New Orleans, Washington, and Philadelphia. Before the extensive construction of railroads even ports such as New Bern, North Carolina, whose entire county listed fewer than 20,000 people, were sailed via Hell Gate.

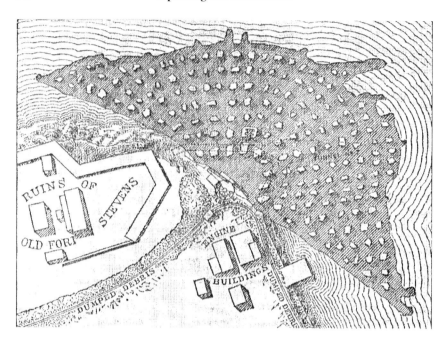

**Though printed at a later date, this aerial view Newton's work site published as part of a summary of General Newton's work. Inside each of the galleries, the white squares represent the piers holding up the roof of the tunnels (*Popular Science Monthly*, February 1886).**

Though the $225,000 appropriation did provide some funds for Newton to work with, it was but a third of what he asked for, and the shortfall led to being able to hire just 125 workmen rather than 250. Because he did not have enough laborers, he had to assign the men he had on duty to lesser amounts of work. Just because the funding wasn't here did not stop commerce. So shipping on the river did not halt and just on July 3, 1873, one newspaper counted ships passing through Hell Gate at 13 who went south and 35 who sailed east.

The workmen seeing Hallet's almost like a cave led to the production of this engraving of the site. Above the tunnels can be seen stretching under the river and the square supports for the roof of all the tunnels.

## Newton's Additional Duties on the Hudson as the Economy Crashes

As chief engineer, Newton was also responsible in the time beginning on July 1, 1873, for some military jobs, including artillery emplacements at

Fort Columbus, Castle William, Fort Wood, and Fort Hamilton, all in the New York City area. Also, among his assignments were reporting on Fort Montgomery on Lake Champlain, Galveston Harbor, and at Otter Creek, Vermont, among a dozen others. Newton supervised the free passage on the Hudson which was slowed where the river narrows at Shad Island (south of Castelton-On-Hudson) and then further north at Schodack Island near New Baltimore. Finally, north of Albany, Newton was responsible for ships' safety at Green Island, opposite the city of Troy. Back in the city, the General was being called on for his expertise by the Department of Docks throughout his tenure with the Corps of Engineers for reports on the conditions of piers and bulkhead walls. These sites included the one at Canal Street and, among others, river walls on the Hudson at King Street.

To help alleviate some of the currents at Hell Gate, on July 18, 1873, Newton asked for money for work on the Harlem River whose speedy flow, four knots, into the East River severely complicated navigation. The combination of the tides moving back and forth, and those forces twisting the East River's flow by rocks and reefs caused unreliable navigation whether under steam or sail and whether the ships on their own or subject to the pull and slack of a tug rope called a hawser.

Under the direction of Lieutenant-Colonel and Brevet Major-General Newton, of the Corps of Engineers the Lower Harlem River was surveyed in the period from March 19, 1873, to February 21, 1874. During the next two years he would order a survey of the Upper Harlem River near 225th Street, a study lasting from July 2, 1874, to February 10, 1876.

And as work continued up and down the Hudson and in the harbor on the New Jersey shore, and as the size of the ships entering the harbor grew, so did the needs of the city and the responsibilities of Chief Engineer Newton. One of his assistants during this time included Lieutenant W.H. Heuer, who had been a vital participant in the lowering of Blossom Rock in San Francisco.

In the fall of 1873, the railroad-investment bubble burst. It had sustained the national economy in the years after the Civil War. Although much was accomplished by Newton that year, 1873 was marked by the "Panic of 1873." On September 20 stock market trading was delayed and then a suspension on trading was ordered and stayed in effect for 10 days. Nearly 300 brokerage houses went bust. This event began a period known as the Long Depression—it lasted for six years until 1879.

The collapse of the real estate market in New York City quickly followed. A wave of bankruptcies overwhelmed investment houses and other businesses. By the winter of 1873–74, a quarter of New York's labor force was out of work and wages had declined for many who kept their jobs. Workers, skilled and not, were affected at Hallet's as well.

For Newton's work on the rivers in New York City, the depression that began on September 20 meant that on November 20, 1873, much work would be discontinued until August 1874.

Meanwhile, the general's family grew. He had married Anna Morgan Starr, of New London, Connecticut, in 1848. The couple had 11 children, 5 of whom would live to adulthood. They welcomed into their family a girl, born in 1873, probably in March. Named Mary Anne, this surviving child would be the focal point of two very large explosions in just a few years.

## Advantages and Dangers of Nitroglycerine

By 1873 Newton now knew much about explosives for underwater work. He knew that once the nitroglycerine was installed into drilled holes, and the tunnels and galleries then flooded with the East River water, the resulting combination of water and explosive amplified each other. If those in charge of appropriations couldn't see it, on November 9, 1872, the *New York Herald* had congratulated Newton for apparently solving the removal of Hallet's Point, "a problem that has defied previous attempts and is itself a pioneer in a new method."

But for readers of the *New York Times* problems remained with explosives: "Another Nitroglycerine Explosion," February 6, 1873. Near Tidioute, Pennsylvania, a man, working with nitroglycerine in his own house, not at a site, blew himself "to atoms killing his wife and fatally injuring his child." "Nitroglycerine Explosion," the *Times* headlined months later, July 7, 1873. At the works of the Cambria Coal and Iron Company of Pottsville, Pennsylvania, the supervisors had chosen to store gunpowder and nitroglycerine in a shed. Some townspeople, it is assumed, looking for explosives to be used in celebration of the Fourth of July, went into the shed. The explosion was first thought to be an earthquake. "If there were any persons killed, it would be next to impossible to find their remains amid the terrible upheaval of dirt and stones."

General Newton was not alarmed about handling nitroglycerine, particularly the need for constant coldness of the oily liquid. The science he had read about and the experiments reported on from Willet's Point satisfied him. His own experience and reading showed that the explosive's essential chemicals could be transported, combined on the scow or elsewhere, and used safely. He was aware of the precautions and made certain that his subordinates and his workers understood all those warnings. The science did not mean that some delays weren't inevitable, for much of the preliminary work on the river was still outstanding.

## *Divers and the Positioning of Explosives*

In one of his reports, Newton insisted that "The bottom must be explored by divers, loose rock, and hard pan removed, and soundings taken before a position for the dome [of the scow] can be selected. For these reasons operations have been intermittent here."[1] Added to that was the fact that the complete use of fully tested breakthroughs in diving were only just then being accepted. (In Maillefert's time, the early 1850s, an iron rod from a rowboat was used to determine the depth and general characteristics of the river's bottom.)

In the judgment of an issue of the 1872 *The Century*, "Weighted down with heavy armor, the diver's task requires the judgment of an educated miner, as well as the skill of a diver."[2] Even when the divers had done their best, their reports were often so "vague and conflicting, that the superintendent of the scow [Striedinger] was at last compelled to add diving to his other accomplishments, and do his own submarine surveying."[3]

Apparently, there were two massive devices not yet in service but included in that original January 1867 estimate by Newton: one iron floating current-breaker, 160 feet long, and one iron floating current-breaker, 100 feet long. The method to aid the divers had to be worked out, for moving and placing these two massive current breakers into place would not be easy.

But by 1873 a modus operandi was developed for the East River work due in part to the inventions and advancements in the tools a diver needed and due in part to those current breakers which wrapped around the underwater work area. In previous times, to work for short periods in locales like the East River, underwater work was attempted using diving bells, but the bells were limited in the depth they could reach. To have a diver be able to work at great depth in safety and with efficiency, four things were needed: the diver had to breathe, to see, to be protected from the cold of the water, and to move about with some ease.

Two advances in diving equipment came about in a five year period. Augustus Siebe's modification to the Deane Brothers' 1834 helmet led to a breathable helmet. In that same time, Charles Goodyear mastered vulcanization, a process by which the driving suit could be made waterproof and flexible. So, a diver could now see through the window in the helmet and also receive air from a pump on the surface. The suit protected him and the use of weights gave the worker the ability to move on the river bottom. The darkness in the water became manageable when Joseph Cabirol invented an oil lamp that would work underwater and so it could be said that by 1860 all the requirements were in place. True, work needed still be done on a rebreather to extract the carbon dioxide that the diver produced inside the helmet, but by the time Newton needed the divers, many of the problems of three decades before had been solved.

The diver on the East River was probably hoisted over the side of the scow, his suit being so cumbersome out of the water. Once underwater, his first job was to reconnoiter the rock to be worked on and look for a workable area onto which the dome might be lowered. There exists no information about how the diver talked to the scow, but all evidence shows no true communication until years after Newton's tenure in New York. By tugging on the line, he might at least signal that he was finished inspecting the rock and was ready to be hoisted back onto the scow. Once the diver was being made ready, a rowboat was sent to collect the explosive to be used that day. (One source for the nitroglycerine at the 1873 stage in the work might have been nine miles away in Communipaw, New Jersey, more specifically Communipaw dock number 7 in Jersey City Harbor.)

Back on the scow, once his helmet was unscrewed, the diver could help direct the scow's position and the vessel could be moved, secured with chains and anchors, and the drilling dome lowered until its adjustable legs were solidly stable. A steam tug might be called on to maintain the scow's steadiness in the current.

Next, the holes would be dug into the rock originally using the hammer-like drills inside the drilling tubes of the dome, and the dome withdrawn. *Scribner's* magazine describes how "the rock is pierced by the dropping stroke of the heavy drill-bars, falling sixteen inches clear, and penetrating from six inches to two feet an hour, according to the hardness of the rock."[4] Suited up once more, the diver might descend to check on the work, see to the condition of the drills, all the while being protected from currents by using the dome as a kind of shield.

With the drill work completed, the diver descended again to secure the drill holes from any silt that might accumulate in them. This task is necessary because the underwater work had proven to be so exhausting, and the tides and currents so swift that the drilling and the exploding must be separate undertakings. Using corks with lines attached, the diver tightly stuffed the holes, pulled on the air hose or rope and was hoisted aboard. There may be more than one diver working at a time.

At a suitable time, perhaps the next day, the nitroglycerine having arrived onto Mill Rock in cans, the workers transferred the liquid into cartridges to be inserted into the drill holes. A small number of cartridges are then taken by rowboat to the scow and are made ready for detonation (wiring, primer, etc.). The diver is hoisted and lowered—we do not know if the cartridges were in his grasp or not—and he begins to find the plugs by finding the lines attached to the corks. He inserts each cartridge in each hole and is lifted back up.

The weight, the bulk, of the scow, might then be positioned, by Newton's design, to absorb most of the shock of the explosion. A scow 127 feet long, 58 feet wide, and weighing 750 gross tons (1,680,000 pounds) certainly would absorb much of the shocks. (Just one more part of the genius of Newton's design.)

Striedinger writes of the blast's impact being "imparted to the bottom and the sides of the scow," and so the effect on surrounding vessels and property was "very much lessened by now acting upon a vessel freely riding at anchor. The careful execution of this rule enabled us to fire large blasts at a comparatively short distance without the slightest inconvenience…. For instance, not the least jar was felt on board the scow when 1,003.5 pounds of nitroglycerine, contained in 10 drill holes, was touched off at a distance of 275 feet."[5]

Back in, 1871, operations had begun on Coenties Reef. Reports on this work differ, including the range of nitroglycerine used; one version claimed 17,137 pounds and another 22,606 pounds with 400 charged holes and 56 surface-blasts. The reef would not be lowered to 25.5 feet until as late as the spring of 1876.

When a model of the Hallet's Point works was sent to the Vienna World Exposition (May 1, 1873–November 1, 1873), the exhibition of the Hell Gate blasting systems earned Newton's assistant Striedinger a gold medal. A similar model and description of the apparatus used by him for the ignition of mines caused the King of Spain to decorate Striedinger with an order of merit.

## Changes and Adaptations in American Shipping

The year 1874 marked an attempt to revivify America's steamship industry. By industry, it must be understood that the word included not only the building of steamships, but also the reclaiming of a bigger share of the shipping business. The diminished number of American steamships meant that many American goods would be carried all over the world in non–American vessels. Great Britain had leaped ahead of the United States in the production of steamships, partly because of the American Civil War. As early as 1857, the year of the Panic of 1857, too many clippers lay idle at piers and docks in Manhattan and Brooklyn in the midst of a national recession. (The last American clipper ship was named *Glory of the Seas* and launched in 1869.)

While wooden sailing crafts were limited in production, nevertheless, many that were built previously in a 25 year period accounted for a great deal of tonnage. And more than 300 "Down Easters," large wooden sailing ships, were built, mainly in Maine for coastal trade, built just as the clipper ship builders halted making their swifter vessels. While the ships built in Maine sailed the routes to California, those remaining clipper ships, carrying lighter loads, sailed long distances such as New York to Hong Kong.

Even so, steamships were the future, particularly with the invention and improvement of the screw propeller to replace the paddlewheel. In 1849, the federal government called for bids for a subsidized mail service contract between New York and Liverpool, specifically to compete with the Cunard line of Britain, which was financed by the British government beginning in 1840.

The American contract was awarded to a company later to be referred to as "the Collins Line" after its owners. For contrast, we know that the sleek clippers of the 1830s and 1840 might measure about 800 gross tons. The four ships of the Collins line—the *Atlantic*, *Arctic*, *Baltic*, and *Pacific*— were not only 2,900 tons but one-sixth larger than the biggest Cunard ship. In 1957, *American Heritage* remarked

> Service started with the Atlantic on April 27, 1850…. The Atlantic justified Collins' expectations by breaking Cunard's westbound record on her return trip, and on her second voyage she broke all records both ways. In April 1851, the Pacific made history with the first crossing under ten days. But the ships of the Collins line were ill-starred, and the line went out of business in 1857.[6]

The government did not give up however and in 1865 the Pacific Mail Steamship Company was awarded a $500,000 annual subsidy—later one million dollars—to operate a steam packet (i.e., a ship carrying passengers, small packages and mail) between the United States, China and Japan. The government decided to demand a ship of at least 5,000 tons, built in the United States.

To give a sense of the value of the ship, later it was reported the vessel arrived after 26 days at sea from Hong Kong, a distance of 20,122 miles. This meant the vessel covered 672 miles per day, or 28 miles per hour. "She brought about a dozen cabin passengers, the usual heavy cargo of tea and rice, and 160 Asiatics, as well as 16,246 packages of U.S. mail, 3,350 tons in cargo, $24,700 in treasure."

Perhaps the best way to understand the continued importance of shipping to America and New York is to state that in 1874, nearly 61 percent of all U.S. exports passed through New York Harbor. And the country was developing railroads as well, by one count in 1874, 1,300 of them.

## Hallet's Point Reef Preparation Continues

Meanwhile, the Hallet's Point tunneling continued. Up until June 1872, the digging was accomplished first by hand-drilling and black powder, and afterward by machine-drills and quicker, more efficient explosives, with the carving of tunnels, galleries, and piers lasting from July 1869 to June 1875. The shaft had been excavated from behind the 135-foot coffer dam, from which 41 radial tunnels, long and short, were cut in the rock, with 11 transverse galleries. Supporting the roofs stood 172 piers of natural rock. "The roof was left 20 feet thick in places where the rock was unsound, and elsewhere varied from 6 to 15 feet; whilst the columns averaged 10 feet in thickness and rose from 8 to 22 feet high."[7]

Using the explosives at the tunnels and galleries at Hallet's Point and the rock removed at Coenties Reef, the projects were ridding the river of 18,503,100 pounds of rock.

As the finish point for Hallet's drew nearer, newspapers began to intensify their coverage of the event, looking carefully at all the newer technology Newton had been adopting and creating. In March 17, 1874, the *Brooklyn Daily Eagle* noted, even as the East River Bridge was being constructed "probably no more important engineering enterprise has been undertaken than the removal of the rock obstructions at Hell Gate."

## Newton's Jurisdiction Extends into Vermont

Furthermore, for fiscal year 1872–73 Newton was directed to take on the following tasks: Improvement of Passaic River • completing detailed surveys in New Jersey including Rutherford Park Bars • improving Belleville Bar in order to clear the Passaic River up to the city of Passaic • removing some obstacles in the harbor of Port Chester • on the Hudson opposite Rhinecliff, undertake improvements on Rondout Harbor • survey Keyport Harbor, New Jersey, for a breakwater at Rouse's Point, New York • survey the channel between Staten Island and New Jersey (now called the Arthur Kill) • survey of the Swanton, Vermont, Harbor.

Thus Newton's authority extended for more than 275 miles and over two major rivers, one ocean, numerous harbors, and many creeks.

This is not to say that Newton himself accomplished all of these tasks but they all lay within his responsibility. The multi-talented Julius Striedinger, having studied at the mineralogical laboratory at Yale, was given the additional tasks of examinations and surveys of the East, Harlem, Passaic and Raritan rivers, Keyport and Port Chester harbors, and the principal waterways, for the purposes of improvement.

## Explosives, Fuses and Continuing Dangers

Newton, in charge and displaying his expertise and thoroughness, required that engineer Striedinger experiment "with the fuses offered in the market, to test their ... merits in exploding frozen nitroglycerine." These experiments led to his building better fuses for both "frictional and for voltaic electricity ... safe to handle, impermeable to water, and fully reliable to explode frozen nitroglycerine."[8] Newton was one of the few to use nitroglycerine at the time. Earlier, for example, the construction of Central Park in New York City opted for the use of 166 pounds of dynamite. Central Park would not be completed until 1876.

Searching for a secure spot to store his nitroglycerine, the General selected Flood Rock. More than that, it would be Flood Rock that would be the target of Newton's second great explosion. The obstruction in the river would first serve as a storage area, and later, both a factory and the project unto itself. Because Newton had discovered through the surveys by divers that Flood Rock and Middle Reef (also sometimes referred to as Middle Ground) were separate entities. Middle Reef meant a combination of separate rocks—Hen and Chickens and the two Negro Heads, with some simply observable high points on one group of rocks. At times, the Gridiron was falsely put into that group as well. But Flood Rock, east of the Middle Reef, was the next largest problem in the East River after Hallet's Point.

Newton had concluded that for effectiveness and safety his explosions on the smaller rocks and reefs had to be limited to a maximum charge of about 500 pounds of nitroglycerine and that his workers and their boats must be 600 feet away at the time of the detonation.

It was Striedinger's expertise with detonations plus Newton's scow with its drilling apparatus that enabled the work on rocks in the river to proceed more smoothly. In 1874, Striedinger's duties also made him responsible for reporting on what he considered to be the best practices for the handling of explosives for those obstacles still remaining in the river.

When work could begin again early in the year, the assistant's report to Newton explained the mandated safety precautions currently in use on the steam scow. Perhaps Newton applied the experience of George Mowbray from his work on the Hoosac Tunnel. It appears that his company, Mowbray, Gilbert & Warren, supplied some of the nitroglycerine for Newton.

To carry out Striedinger's ideas for underwater work, four materials or procedures became vital to success. Again, it seems that this preparatory work was completed on Flood Rock, a safe distance from land.

Various adaptations by Striedinger were made depending on the depth and surface of the rock, the problems encountered with inserting cartridges into drilled holes, the choice of fuses, and the amount of nitroglycerine (in pounds) to be used for each explosion. The nitroglycerine was stored on Flood Rock until the chunks of ice in the river interfered with movement between the scow and the rock, necessitating that the nitroglycerine be "carefully secured" on board the scow. There the frozen nitroglycerine was warmed in a large wooden tank, filled with water not exceeding 85° Fahrenheit.

During September 1874, experiments were performed in the search for better fuses. And to secure the watertight fit of the cartridges of the explosive and the primer, a combination of wood, paper, copper, bees-wax, rosin, and tallow. The wires were coated in gutta-percha, a naturally occurring rubbery plastic substance from the early 1840s that had the three qualities of being stiff, inert, and hardy.

The batteries were made ready. In a thorough description of what came next the drilling-scow was withdrawn a "distance of 150 feet for charges below one hundred and seventy-five pounds, and a distance of about 250 feet for charges upward and to five hundred pounds." After the drilling-scow had been removed, "The nitroglycerine, which was the principal material used in these blasts, was brought to the spot on another [smaller] scow, and there filled in tin cases of various lengths, were lowered to the diver."[9]

If it was not possible to place the cartridges in holes or in the case of surface blasting, Striedinger notes "these charges were provided with weights and lines, to secure close contact to the rocky surfaces."

For Newton's workers and staff, the year saw a minimum of 17,000 pounds of nitroglycerine being exploded (along with 39 pounds of dynamite) and yet "no fatal accident happened during the whole season.... Not one misfire happened," Striedinger proudly wrote.[10]

But not everyone was Striedinger and so beyond Hell Gate, many were not so careful. In 1874 in January alone the *National Aegis* of Worcester, Massachusetts, told how while chopping wood, a chip flew through the air and hit a can of nitroglycerine killing three men. A few days later, the *Hartford Daily Courant* reported four men killed at the factory at Communipaw where Newton was buying his explosives. Further south, the *New-Orleans Times* whimsically wrote of a baggage handler who tossed a trunk onto a station floor. Inside one of the pieces of luggage was a can of nitroglycerine and the paper reports of the handler that "there was a funeral in his family on the following day, and he headed the procession." The *Dallas Weekly Herald* said that 12 miles outside of Dallas, while a can of nitroglycerine was being filled, "Mr. William Spots ... was blown to atoms."

And on June 19, a *Times* report from Staten Island gives us the sense of newspapers at the time:

> Walter Hicks put glycerine near the stove to dry, when it suddenly exploded tearing the stove to atoms and casting its fragments about the room. Mrs. Hicks, aged seventy, was thrown by the shock into the pantry senseless, sustaining a compound fracture of the leg near the thigh. It is thought she cannot recover.... Mr. Hicks was severely injured about the hands and arms. ... A man and a horse passing at the time were knocked over.

In June 1875, Nobel's own nitroglycerine factory exploded in Lysaker, Norway. Though the explosive chosen was intended to work effectively depending of the hardness of the rock, sometimes, it seemed, the cost of the explosive determined the choice of its use. It became clear that there Newton tried to avoid a tradeoff between effectiveness and safety of the various explosives.

Care, partly by General Abbot, had been taken to test the various kinds of explosives and detonators. Gun cotton as one of the detonators was described in 1847 as "three parts of sulphuric and one of nitric acid and rough cotton." It had been used as the explosive itself for some time in mining. An 1875 journal, the *Canadian Journal of Science, Literature and History*, concluded that "we may say that nitroglycerine explodes with rather more than six times the force of ordinary blasting powder, and not quite twice the force of an equal weight of gun cotton."[11] (Gun cotton, however, turned out to be far more useful in munitions than in detonations.)

Abbot at Willet's Point set to work on such complexities as electro-motive force, and the resistance of the battery, fuses and wires, and the result was a formula by experimentation. General Newton wrote, "Up to the middle of 1874, nitroglycerine had been principally used for blasting purposes" as well as other "nitroglycerine compounds. Neither of them was found to be as powerful as the nitroglycerine itself; but it was repeatedly demonstrated that, with 10 ounces of Rendrock or Vulcan Powder, they could break as much rock as they formerly did with 8 ounces of nitroglycerine, while the cost per lb. was less than one half that of the nitroglycerine."[12]

Throughout Newton's two decades of work, many different kinds and strengths of explosives would be used. Many of these brand names have been mostly lost to history. With the research and experimentation by General Abbot and with the testing by Newton and Striedinger, some data began appearing.

In the time since Nobel's invention of dynamite and as experience working with dynamite grew, measurements had been made to demonstrate that nitroglycerine explodes at 210 degrees, gunpowder at 600 degrees, but dynamite does not detonate until somewhere between 1,400 and 3,200 degrees. The danger of heat might now be safely discarded.

At last Striedinger wished to praise his fellow engineer, writing

Before closing this report, I beg leave to congratulate you upon your eminently successful design, the United States steam-drilling-scow, assuring you, at the same time, there is no doubt in my mind that the United States steam-drilling-scow, worked by last year's crew of picked men, will successfully remove any submerged rocky obstructions to a greater depth, in a shorter time and at lower cost, than any contracting party.[13]

## *The Newton Steam Drilling Scow at Work*

During the slow, unfunded period of the spring of 1874, the Newton Steam Drilling Scow underwent repairs, and "its dome was provided with a new set of enlarged drill-pipes for allowing the use of larger drills" as Newton let his superiors know.[14] The chief engineer used the time to order some soundings and investigate the character of the tidal currents. Those currents led to two notable disasters on the river. The *Fanny Fenn*, a schooner, sunk in Hell Gate in 20 minutes. Soon after "the Schooner *T Morrell*, carrying ore in 1874, sank with a total loss of $15,000 plus the loss of $8,000 cargo." Even a powerful vessel, the steam tug *Lilly,* was not safe in these waters. The *Lilly* blew up and sank at Hell Gate on November 17.

With the sinkings and losses happening far too often, the New York Ship Owners Association sent agents to the federal House of Representatives, whose pleas for funding were then to be handled by the Committee on Commerce. The pleas may have accomplished some good because Congress appropriated $250,000 for 1874–1875.

Newton remained busy even in the midst of a suit for patent infringement brought by Samuel Lewis, focusing on the scow, the dome and the drill guides. By 1874, the court, in a 21 page decision, decided

> General Newton appears to have considered Lewis' plan, and to have deliberately rejected it, and to have proceeded on one directly opposite…. He took up the apparatus where Lewis left it, and discarded Lewis' arrangement. These views are sustained by the experts for the defendants, General Tower and Professor Peck, and by the other evidence in the case. A decree will be entered dismissing the bill, with costs.[15]

At the same time Newton and his staff and workers, while finishing up the necessaries for Hallet's Point, set to work on needed river and harbor work at Albany and Troy. Work continued dredging mud bars in the Hudson, surveying reefs, taking soundings, and measuring currents in New York Harbor. So much debris traveled into the harbor, both on the New York and New Jersey side of the Hudson, that the depth of the river and the harbor increased even with dredging. This fact continued to underline the necessity for more navigable waters in the East River.

Thus, the removal of Hallet's Point became so vital. The reef in the shape of a semi-ellipse, extended 720 feet in length along the shore, and to 300 feet in breadth out into the channel. To be removed, to secure a depth of 20 feet at mean low water, amounted to 53,971 cubic yards. The reef was dangerous, not only in itself, but also on account of the eddies caused by the tidal currents on either side of it, according to their direction.

As for the smaller obstructions, part of the remainder of General Newton's scheme, the continued lowering of those rocks resulted in a decrease in the speed of the current and the number and power of the swirling eddies. One such obstruction was Way's Reef, north of Pot Cove, and on August 8, 1874, the work gangs started to attack. It may be that the workers on the scow were ferried to work by the tugs that moved the scow from rock to reef. The day-gang, with 30 workers, reported for work at 7:30 a.m. and continued until 4:30 p.m.

As Superintendent Striedinger counted, "The average working force of the United States steam drilling-scow, while engaged on Way's Reef, numbered thirty-seven men.... The night-gang consisted of one diver and four sailors, making submarine surveys, and marking, with weights, the points for drilling; if drilling continued at night."[16]

## Changes in Newton's Plan

In 1874 the Hell Gate project was modified, first by omitting the construction of the sea walls and dikes, by including Diamond and Coenties Reefs in the estimate, and by reducing the estimated cost of improvement. This reduction was due to lessened unit costs resulting from Newton's improved methods and purchase of tools. Using the diving equipment, the cost of lowering these rocks by surface drilling, blasting, and removing debris was lowered to about 51 percent of the original budget. The multiplier of 22 is best applied again to get a sense of what the projects would cost: in the year 2020 is more than $113 billion.

Experimentation and testing continued on explosives with lesser-known brand names—Ajax, Vigorite, Rippite, and Earthquake Powder. As time passed the ones favored, depending on the density of the rock, turned out to be gunpowder, dynamite, nitroglycerine, and Rendrock (soon to be call Rackarock). Newton reported that by 1874 there had been 47,401 of cubic yards rock removed from Hallet's tunnels.

In 1875, down and up the river and even into the harbor itself, the further work of lowering the rocks and reefs continued such as Coenties and Diamond Reefs to the 26 foot level, in those years a depth thought to be sufficient to accommodate any vessel. On the scow, the platform for the blasting of the rocks in mid-river, Newton insisted on precision in each placement of the dome describing how "during the drilling-operations, by means of sextant-observations ... positions were plotted ... and finally transferred upon the drill-hole sheet, upon which each separate hole as actually drilled, charged, and exploded, was carefully laid down."[17]

Had there been any doubt, any ignorance, about the importance of Newton's work on the East River, and not just to the city, an officer of the

Captioned "Sectional View of a Transverse Avenue," this image allows a sense of the height and width of the tunnels and galleries. Still in use are sledges and pick axes for some of the finer work (*Frank Leslie's Illustrated Newspaper*, October 7, 1876).

New York Chamber of Commerce, delivered an address on December 7, 1875, pointing out that "…in the fiscal year ended June 30, 1875, the total amount of Customs revenue for all the United States was $157,167,722, of which $109,207,786 was taken at the port of New York."[18] (The ships choosing to travel through the Sound and the river contributed to that significant amount of money, that 69 percent of customs revenue.) Including gross public debt, federal spending for that period reached $2,756,000. So customs at the Merchant's Exchange Building at 55 Wall Street alone took in 57 percent of all federal spending for the year.

Since 1874, the following East River dangers to navigation had been blasted and their depths increased to the 26 foot level: Shelldrake, a reef identified only as "near the North Brothers," and the Heel-Tap Rocks, north of Great Mill Rock. Great Mill Rock was not a peril since it was an obstacle easily avoided by shipping. Four major obstacles in the river could be written down as reaching the 26 foot level of depth by 1875, even if the dredging of the blasted rock had not yet been completed. The time was fortuitous since, not for the first time, the East River froze in 1875, negating any work to be done on the water for winter time. Other work continued, when possible, with the assistance of both Lieutenant Heuer and Mr. Striedinger on some reefs as well as at Hallet's Point.

## Two Rocks and a Reef: The Frying Pan, Way's Reef, and Pot Rock

General Newton decided to attack the Frying Pan and Pot Rock, both near Hallet's and Ward's Island, but closer to the Sound. (The names of Pot Rock and Pot Cove indicated that somewhere in that area was swirling the whirlpool referred to as "The Pot.") Using a tug and the vessel constructed out of the marvelous amalgamation of the brain of Newton with later adaptations by both Newton and his superintendent Sidney S. Shelburne, ongoing work at the Frying Pan and Pot Rock yielded up to 6,119 tons using 5,479 pounds of nitroglycerine.

Way's Reef, with an original depth of five feet, had been lowered by M. Maillefert in 1851 to 17.2 feet. Newton was obliged to continue. By the General's statistics, the reef measured 235 feet long by 115 feet wide. Described as being found near Woolsey's Bath House (near Pot Cove), its 6,200 square feet meant that 44,200 cubic yards of rock must be demolished. This estimate of the size of the work was discovered by:

> a systematic survey and resurvey of the whole reef, by means of divers. From 12.15 p.m. August 4, 1874, to 4.30 p.m. January 20, 1875, 134 hours were needed to finish lowering Way's Reef. Striedinger's precise report describing the work. continued the scientific aspects of Newton's work, while giving strict accountability regarding the money spent on this project. So we read that "262 holes were drilled using 4 to 5.5 inch bits, gouging out 3 to 12-foot holes into which cartridges of explosives could be placed" and "65 drill-blasts and 16 surface-blasts were made, and 16,792.3 pounds of nitroglycerine and 381 pounds of dynamite were consumed. Loose stone was either hoisted on deck or rolled into deeper water by the divers."[19]

Owing, first, to a blunder in the title of the appropriation bill, the dredging tasks were obliged to be delayed. In addition, because of the year's frozen river, the drilling scow and other floating property were laid up alongside the dike between Great and Little Mill rocks during the winter. This time allowed Newton to reflect that the work of preparation for the explosion of Hallet's Reef which might have been completed in four years, had by then extended over six years and ten months.

## Spuyten Duyvel and the Harlem River

Now falling under his command was the improvement of the harbor of New York with regard to the opening of Spuyten Duyvel Creek, a troublesome body of water that runs northwest into the Hudson and flows as an extension of the Harlem River, the connecting body of water southeast to the East River and thus into Long Island Sound. In 1826 and again in 1863 attempts were made to tame the treacherous and narrow creek. The length of the waterway from the North River (Hudson) to Little Hell-Gate, measured through the Spuyten Duyvil Creek and the Harlem River, is about 39,000 feet, nearly eight miles.

With the need being expressed for the two waterways being made navigable, "At a small cost," Newton wrote, "in comparison with the accruing benefit, a channel can be made from the North River to Long Island Sound, through the Harlem River, with greater depth of water than the North River affords at some points between this city and Albany, and of width sufficient for all practical purposes of the commerce that will seek to use it."[20]

And yet, 30 years would pass before even the first section of the project, to be called the Harlem River Shipping Canal, was completed. In proportion as the survey advanced, additional tide-gauges were set up near McComb's Dam and the High Bridge. Continuous tidal observations plus soundings were likewise made during 32 consecutive hours.

Through the numerous monitoring stations along the shore Newton had learned a great deal, for example, about currents in the East River. When he reported that the "Maximum ebb and flood currents set in nearly opposite directions—Mean rise and fall of tides is in Pot Cove (six feet)," he was directly speaking of the spot nearest Hallet's Point, assured that the current's effect on navigating would be lessened by the destruction of Hallet's.[21] He knew also that by accurately measuring the current and by taking advantage of the data on tides and tidal currents he might more efficiently place the scow.

To give a sense of the traffic on the East River, one typical ship from the Fall River line named *Providence* was 373 feet long and could comfortably care for 840 passengers. On October 15, 1872, Commodore Jim Fisk's famous steamer was going through Hell Gate when she collided with a drilling machine and was badly damaged. Her passengers were transferred to the steamboat *Stonington*. We can only imagine the panic among passengers, the expense to the owners, the delay, and the chaos on the river due to this accident.

It was not only the river that caused problems. It appears that the steamboat *Stonington* itself developed a terrible reputation, so much so that a collision between it and *Narragansett* was included among the work of Currier & Ives, and may be taken as what might happen until Newton finished his work. It should not be forgotten that these vessels were wooden with fires as part of their motive power.

With Ways Reef being destroyed, Newton could look back at year's end on the lowering to 26 feet of Coenties Reef on August 27, 1875. Many, many examples exist, but the following data from the work in one fiscal year on Coenties Reef allow a flavor: Number of holes drilled 126 • Number of feet drilled 990 • Nitroglycerine used: for 20 drill-hole blasts, 6,037 pounds • for two surface-blasts, 279 pounds • amount of rock removed by grapple, 654 cubic yards.

Work was completed on the Frying Pan, and much of Diamond Reef by November 24, 1875. After blasting, the dredging could begin, as soon as Newton approved the contracts with the companies that accomplished the dredging and the companies that moved the scows to carry away the blasted rocks taken up from the bottom. Some of the loaded scows took away the dredged material for building a better seawall at Blackwell's Island.

The careful work being completed along the shoreline at Hallet's Point by now included the erection of four demolition sheds, each built distant from the others, each small building to hold no more than 25 pounds of dynamite.

Ever cautious, Newton had to consider that care needed to be taken to keep the roof of the tunnels at a minimum distance of ten feet of thickness below the riverbed because small blasting and tunneling continued throughout the project. Sometimes, however, the distance shortened to four feet below the surface. Continued observation in the tunnels was necessary due to the leaking of East River water through the roof of the tunnels and galleries.

Marveling at the precision of the work, one newspaper commented that Newton's plan displayed "the rare spectacle of a government work done thoroughly and economically … work of ingenuity and practical skill."[22]

Those numbers above suggest at least two ideas. First, that since he was accountable to the federal government, the General would be sure to demand of himself and his assistants an extraordinarily complete listing of materials. Second, for purposes of advancing the knowledge of anyone in the fields of explosives, demolition, and civil engineering, he would be precise and thorough in his work and in the reporting of the work.

Over the years, Newton either performed himself or had others do it for him in a vast array of circumstances and experiments with explosives. With regard to both safety and efficiency, he had learned about storage, about transport, about the most secure temperature in handling the explosive. He had learned about fuses and detonators. He had learned about drills, and steam engines and the best kind of craft to have on the river to best finish the destruction of obstructions.

Some of this knowledge came from his own engineering experience and the experience of others he had hired. Some came from knowing with which manufacturers to deal. Some came from scientific magazines and reports from others in the Corps of Engineers. Just as many kinds of explosive compounds were tried, so was it true that many detonators were used, depending on the job at hand. Sometimes fulminate of mercury was selected as the detonator, a substance than could be set off by shock or heat, the heat caused by a wire inserted into a cartridge containing the various explosives.

Newton, ever thorough, ever careful, ever thrifty, had experiments performed during the progress of the work. Those trials proved that as much rock could be broken with 10 ounces of Rendrock or Vulcan powder as with 8 ounces of nitroglycerine. And their cost was less than half that of nitroglycerine; also, the use of the pure liquid explosive was less convenient than its solid compounds.

The shaft, tunnels, and galleries were the major work assigned to General Newton, assisted by two others from the Corps of Engineers: Captain W.H. Heuer and Lieutenant J. Willard. In addition to the supervision of Hallet's, Newton's scientific reporting concerning tides and currents, concerning depths, and sailing condition went on. In and around New York Harbor saw the improvement of East Chester Creek, and the harbors of Port Chester and Rondout, a place 100 miles up the Hudson near Kingston, New York. Small projects might be started on the Harlem River and possible projects needed to be studied on the Passaic River and the channel between Staten Island and New Jersey. Work was needed as far north from New York Harbor as Lake Champlain. In these places surveys need to be made, dredging and the repairing of dikes, and tree planting to hold soil in place near waters.

Newton, now an internationally respected engineer, was sometimes sent elsewhere. For example, in 1875 he was assigned to Montreal, where he was asked to develop a "'comprehensive plan' for expanding the port facilities."

## Up and Down the River

It is possible to trace additional activity of Newton's steam drilling scow in 1875. The scow was towed north of Hell Gate, off 125th Street in Manhattan, where an unnamed, probably small, reef was drilled and blasted to 14 feet from August 27, 1875, to October 7, 1875. The next day, October 8, 1875, the scow began work on Diamond Reef (between Governor's Island and the Battery) and stopped its work on November 24. Work on a part of this reef of 93,330 square feet had to be suspended for want of funds and that reef between Governor's Island and the Battery would needed to be attacked for years to come.

Smaller jobs included the small rocks known as Scaly Rock (Astoria), Blackwell's Rock, and the Rock off Woolsey's Bath House near the Middle Ground. At some point, Newton oversaw the installation of sea walls upon Hog's Back, 900 feet from Hallet's point near Ward's Island, and Holmes Rock, off the south end of Randall's Island.

Newton was able to solve two problems at once. Problem one was what to do with all the loose rocks that have been drilled and blasted and then dredged. Problem two, how to negate some of the swirling waters around the rocks and reefs in Hell Gate. The General realized that two rocks called at the time Great Mill Rock and Little Mill Rock might be made into one large rock. Using a kind of breakwater edifice called a rip-rap dike, he sent the dredged stone and rock to be dumped between the two thereby creating Mill Rock which can be found about 1,200 feet from Hallet's Point on today's maps.

Improvements continued on the scow. Added was a Clayton duplex air compressor allowing four divers to work at once, a great advantage over the system of pumps worked by hand, both in economy and the quality of air. Another new tool, a double-end 10 by 12 cylinder hoisting engine, was substituted for the old one.

Newton, as usual, was kept busy with projects around the rivers and the harbor and his reports covered many facets of his varied tasks. The report itself seems a model of diligence, showing Newton's control over the great extent of his command. Always precise, always looking for more efficient and cost-saving ways of drilling and blasting, he never hesitated to experiment, understanding as he did his responsibility not only to his army superiors but to the practice of civil engineering in general. In this way, he extended human learning. If a new engineering tool came along, Newton considered it. For example, in 1875 Nobel invented blasting gelatin (gelignite), a kind of dynamite with an adsorbent base of potassium nitrate and wood pulp. It appears that this substance did not become available to mining engineers until 1886.

By the second half of 1875, local newspapers began to understand how vital Newton's work at Hallet's had become. They also saw the difficulty of working with new technology and new explosives. Before very long, the site in Astoria would see many reporters from newspapers and periodicals, both popular and scientific.

## Close to the Finish

From the *New York Herald*, June 22, 1875:

Marking the completion of tunnels and galleries at Hallet's Point, Newton's bookkeeper recorded the 49,516 of cubic feet hauled up the shaft and put onto barges. After the explosion, no less than 30,000 cubic yards of broken stone will be left under water, all of which will have to be removed by dredging. The excavation now being nearly finished, the matter of caution in blowing up the whole of Hallet's Point began to occupy the minds of the engineers, with not much space before reaching Ward's and Randall's Islands and Astoria.

At Hallet's, the total length of tunnels and galleries then amounted to 6,780.67 feet. There were 172 of them, pierced with about 4,000 drill holes. The roof in the tunnels varied from 6 to 16 feet in thickness.

So, in June 1875, the work of drilling holes in

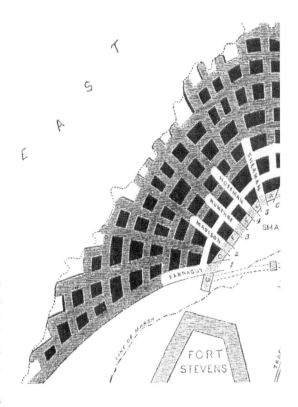

Newton named the Astoria tunnels after engineers, presidents, generals, and admirals (Wikisource).

the roof and piers, to be afterward charged with explosives, was begun. To detonate thoroughly the three acre reef, much study over the years had been taken up, mainly by Mr. Striedinger, General Abbot and General Newton. How exactly would the explosion take place, where would the explosives be inserted, how many holes would be needed in which to insert those compounds of nitroglycerine, and how might the detonation be accomplished with very little disturbance. To be considered first was damage to the river's islands and its buildings and residents, Second, to the buildings and residents of Astoria itself. Third, the area of Manhattan nearest the explosion, around 100th Street, had to be made secure.

As he adjudged that the tunneling at Hallet's was completed, Newton sent some of the drillers over to Flood Rock on June 7, 1875, to begin the first shaft there. Thinking things through as usual for him, Newton had the workers sink "a shaft deeper than we wanted to go to make a well to receive the drainage."[23]

**Demonstrating the breadth of the coverage of the work at Hallet's Point, this image shows workers insert the explosive cartridges in the roof (or ceiling of the tunnels). We also see the flaming head lamps of the workers even as they work around combustible material (*London News*, October 1876).**

At Hallet's Point, drilling the roof and piers so as to make a place to insert the explosives for the final blast can be seen in the engraving above from the *London News*, showing the interest to be international. The white smoke comes from the lights on the workers' hats.

The benefits of the upcoming detonation became much clearer to everyone. The destruction of a portion of Hallet's will widen the passage by doubling it and also have the effect of lessening the speed of the current as well. As a ship navigates on the East River, it reaches a point where the width of the river at Astoria is no more than 300 feet. If more than one ship attempted to pass through at the same time, the danger is doubled. This point, in addition to being narrow, also marks the most difficult passage because of the fluctuating tides, the swirling currents, and unpredictable eddies. The danger here was that on both sides of the 300-foot passage lay Gridiron Rock and Flood Rock on the west and Hallet's reef on the east. Ships piloted cautiously to avoid Hallet's Point Reef had been tossed about and smashed against any of these obstructions. As a case in point, on November 28, 1871, a tug towing a canal boat loaded with coal was run into by a schooner, sinking the tug in four fathoms and setting the barge adrift. That same day the British brig *Alfaretta* ran ashore on the Gridiron, which opening a hole in its side, so that it filled in five minutes. Remembering the narrowness of all three channels at Hell Gate, marine reporters knew the wreck would block the passage temporarily as well, perhaps with ships delaying leaving port and backing up into The Sound.

In the concluding portion of the Hallet's work there were 67 men employed: three foremen, 17 miners and chargers, and 47 helpers. The engineers in charge were identified as: John Newton, lieutenant-colonel of engineers, brevet major-general; James Mercur, captain of engineers; Joseph H. Willard, first-lieutenant of engineers; Julius H. Striedinger, civil engineer, assistant; Bernard F. Boyle, mining engineer, overseer; James Quigley and Robert S. Burnett, assistants. Captain William H. Heuer was for a long time resident engineer at Hallet's Point.

# NINE

# Newton's Many Plans

## *The Country Is Attentive*

Characteristic of the fascination with Newton's works at Hallet's Point Reef can be found in *Scientific American*: "A very widespread interest has been centered on the operations; and the work is one of national importance, although this city is of course more interested in it than any other section of the country."[1]

As part of a celebration of the centennial of the founding of the United States, the "International Exhibition of Arts, Manufactures, and Products of the Soil and Mine"—a kind of world's fair—opened in Philadelphia on May 10, 1876. Featured were a monorail and the telephone. More than 200 buildings were built on the grounds for visitors numbering ten million.

In the fair's category of "engineering works and appliances," visitors could view two products from the mind of General John Newton. Worthy of inclusion among the centennial's other exhibits was a "Model of steam drilling scow, as designed and constructed by Bvt. Maj. Gen. John Newton. Scale 1 to 24."[2] Understood to be equally remarkable was a topographical model of the Hallet's Point coffer dam, shaft, and tunneling at Hell Gate. In the year 1876 Newton was elected to membership in the very prestigious National Academy of Sciences.

Ten years had then passed since Newton reported to his new duty station and assignment in New York City. Many of the obstacle rocks and reefs in or near Hell Gate or close to the Battery, like much of Diamond Reef, had been blasted close to the requisite 26 feet at mean low water. Work too had begun on what was referred to as Flood Rock. The other Middle Reef Rocks—Hen and Chickens Reef, Negro Head, The Gridiron, and Little Negro Head—could be handled by the drilling scow or by some sort of defensive device—at times called a fender—wrapped around the rock.

The staff for this job included the prominent supervisors selected by Newton:

Capt. William H. Heuer, Corps of Engineers, had personal "superintendence of the works from June 4, 1872, until relieved from duty on July 20, 1876."

He was replaced by Captain James Mercur, who had been assigned to study the suitability of various forts in the New York City area.

Mr. Julius H. Striedinger, assistant engineer in charge of the steam-drilling scow; called the "glycerine man."

Lieutenant Willard

Mr. Bernard F. Boyle, the mine superintendent.

Mr. Chester, the electrician.

William Preuss, the surveyor.

By August 1876, printed reports say that the money spent since 1868 to cover the money appropriated for Newton's command equaled $1,340,000, or $29,600,000 in 2020 money. Estimates for the remainder of the work amounted to $5,139,120.

## Hallet's Point Reef

Newton's task, all along, had been to master Hallet's Point Reef, an area about three acres wide, or 130,680 square feet, land that extended 720 feet along the shore and about 300 feet into the Eastern Channel. The width of the reef does not say all about what needed to be demolished, because the acreage was only a measure of the surface of the dangerous obstacle. Hallet's Reef was part of—attached to—the mainland of Astoria. The demolition was, in the restricted language of the time, "submarine."

Newton's scheme had been partly influenced by the work from 1866 to 1870 at Blossom Rock in San Francisco Harbor, but that California project dealt with an obstruction one third the size of Hallet's. The operation at San Francisco involved three parts: a coffer dam to protect the works, a shaft to get down below water level, and then a "chamber" into the rock itself and plant explosives. At the beginning, the tools available were a pick, a gad (a pointed tool, such as a spike or chisel), a sledge, and gunpowder. As the years passed since his arrival in New York in 1866, Newton had access to much more powerful and efficient tools, as those tools were invented and made more powerful and portable.

The history of the Hallet's Point mine, fully illustrated, appears throughout 1876 in the *Scientific American* including remarks on "the comparative cost of hand and machine drilling by compressed air to be 95 cents per foot for the former, and 36 to 37 cents for the latter."[3]

Now, the press begins to concentrate on the layout of Hallet's Point

Reef construction site. Surrounding and protecting the main shaft was the coffer dam. Placed at the top of the main shaft was the derrick to hoist debris out and lower supplies in. Supplying power to the drills and other equipment, the generators were located in the engine houses. The black spread of debris accumulated in two separate spots. Though never fully defined, there seemed to be some sort of railed transport to carry the mined rock, rock extracted from 28 feet below the waterline.

As the surface that was visible above water, of the nine acre obstacle called Flood Rock, was small, the rubble from the Hallet's excavations and some debris from rocks in the river were taken over to Flood Rock and distributed to create a much larger area above water, some 2,600 square yards, where machinery would be placed.

General Newton knew his work had not yet reached completion in the East River. He admitted "Hell Gate is a narrow and tortuous channel ... obstructed by numerous reefs ... rendered much more dangerous by the violence of the abnormal currents or eddies ... [due to] to the rise and fall of the tide in the broader bay below." Specifically, "Hallet's Point ... form[s] the sharpest bend in [one] channel ... here being narrow and crooked."[4] The detonating of the sharp rocks of Hallet's Point Reef would expand the actual space of navigable water from 600 feet to 1,200 feet, though the channel's width would stay the same. It may be that the greatest degree of safety would come about by ships not having to steer clear of Hallet's.

The damage brought about by Hallet's Point Reef, like a collapse on a major road for the flow of commerce, was underscored again in 1876. The schooner *Hippograph* which sank on September 28 of that year was valued at $8,000—money paid merely for the ship's materials that could be pulled from the wreck. The expense of demolishing or towing the wreck was not calculated, just as the cost of having that part of the river blocked for a period of time remained unknown.

Money continued to flow on the New York City waterfront, and the busyness of the port was a source of pride. "The harbour, studded with vessels of every rig and nationality; the great white river steamers, rushing to and fro [and] frequently may be seen one of the great mail steamers, or a three-masted schooner ploughing along under a cloud of canvas."[5] One source measured the annual value of foreign and domestic vessels that sailed the perilous East River with their cargoes at one billion dollars, or $2,739,726 daily. Looking to the future, General Newton was able to estimate that "the amount of commerce and navigation benefited by the completion of this work would be about $4,000,000 daily" in the safety of vessels, cargo and the sale of the items once in port.[6] At the port of New York City, annual custom revenue collected added up to $102,352,037—in

2020 money more than two billion dollars. Every day, on average, brought $280,416 into government coffers.

Also, while it remained true that Hell Gate's perils were many, it cannot be forgotten that a captain's or pilot's option of The Narrows entrance and exit meant different but expensive hazards. That entrance from the ocean might have to be the choice of vessels coming from the south, even though the entrance at Sandy Hook would not be deep enough for another 30 years. The negative aspects for many of those ships included the delay of having to wait for the tide to lift those vessels over the bar at Sandy Hook or the delay and expense of transferring cargo onto or from those barges called lighters.

Given the choice of a wide passage through the Sound, the risks at Hell Gate still was the best choice for many skippers, pilots, and ship owners.

## A Decision on Explosives

The proximity of the Hallet's Point reef to Astoria and to charity institutions and prisons on both Ward's Island and Blackwell's Island (separate bodies then) made it necessary for Newton to devise a system of an explosion which would do no damage to life and property. Years before, the new dynamite—a combination primarily of nitroglycerine and some absorbent inert substance—was crudely tested for safety for the *New York Commercial* in August of 1868, the same year Nobel was granted a patent in the United States. Interested parties tried three experiments with dynamite:

1. a box of 10 pounds placed into a fire—nothing except the box burned.
2. 10 pounds of dynamite without a box—no explosion.
3. a concussion test dropped a 10-pound box of dynamite 60 feet onto rocks—no explosion.

This eight-year-old test proved that only with use of a percussive cap or primer will the dynamite explode. Other, more thorough experiments begun two years previously resulted in several decisions to be made about five aspects of blowing up the reef at Hallet's Point. First, what ought to be the combination of chemicals of each cartridge. Second, how can the fuse, primer, and dynamite in the cartridge be made water-tight. Next, what will be the safest way to set off the explosion of the entire reef, in terms of being certain no detonation may take place until all was ready. Fourth, how to be sure that the explosion will be a contained and effective blast. Lastly, how to keep organized all the 220,000 feet of wires to be connected to both the explosive in the chambers and the batteries on the land at Astoria.

All throughout the process, Newton has been receiving reports from experts like General Abbot, and his staff officers Colonel McFarland and Lieutenant Derby, confirming the safety and ease in the storage of dynamite. Abbot wrote in 1876 that "There was no evidence of deterioration after a storage of five years, the powder being exposed to alternate freezing and thawing."[7]

Another advantage that Newton had was the assignment of Captain James Mercur of the Corps of Engineers who reported for duty August 5, 1876. Mercur's post as adjutant engineer at the battalion at Willet's Point extended from September 14, 1872, to February 20, 1876, and would put him very near the top of all the experiments on explosives that were taking place in those years. He was assigned to superintend the charging of the mines and laying down the electrical system at Hallet's Point.

Another assistant, Lieutenant W.H. Heuer, tested explosives from early in 1875 and remembered the following:

> Heretofore we had been principally using nitroglycerine for our blasting purposes; we then tried several hundred pounds of powder, some Giant Powder [a brand name for dynamite first established March 19, 1868], several thousand pounds of Rendrock, and recently have used substantial amounts of Vulcan Powder. All of these are nitroglycerine compounds. Neither of them is as powerful as the glycerine, but we have here repeatedly demonstrated that with 10 ounces of Rendrock or Vulcan powder we can break as much rock as we formerly did with 8 ounces of nitroglycerine, while the cost per pound is less than one-half that of the glycerine.[8]

(The name "Rendrock" was frequently under attack for patent infringement and disappeared from the American scene very quickly. It had compounds too much like dynamite or Rackarock to be able to claim its uniqueness.) Above all, Newton concentrated on putting together "'just enough power to do the work required and no more."[9]

Newton settled on dynamite, Rendrock, and Vulcan Powder to be used in the cartridges for the final explosion of the cavities and supports below the three acres of Hallet's Reef. (Most of the data on these explosives can only be found in patent lawsuits.) The charging would use three different kinds of explosives, all containing nitroglycerine, the ideal substance because by 1869 a member of the American Society of Civil Engineers could state

> it is universally known that while in powder-blasts the break usually occurs near the centre of gravity of the charge; in nitroglycerine explosions the break is at or below the bottom of the hole, extending outward and downward. Since the force is not directed upward, as we picture a gunpowder explosion, the safety of using nitroglycerine in the middle of the East River between two large population centers stands as a great advantage.[10]

Newton therefore selected the following compounds of nitroglycerine:

1. Rend-rock furnished by J.R. Band & Co.—"can be made to form a semi-pasty mass, which yields to the slightest pressure, and thus can be made to fill up the bore hole entirely."[11] Rendrock powder contained: nitroglycerine, 34.71%; • nitrate of potash, 52.68%; • sulphur, 5.84%; • woody fiber, charcoal, and resin, in nearly equal proportions, 6.77%— 9,127 pounds would be used.

2. Vulcan-powder furnished by R.W. Warren. Cost: 26 cents per pound: nitroglycerine 30.0 %; • charcoal 10.5 %; • sulphur 7.0 %; • nitrate of soda 52.5 %—11,853 pounds would be used.

3. Giant-powder No. 1, or dynamite, furnished by Atlantic Giant-Powder Company, Messrs. Varney & Doe. Cost: 60 cents per pound (this company had an office at 152 Chambers Street in the city): nitroglycerine 75.0 % • kieselguhr 25.0 %—28,935 pounds would be used.

## Making the Cartridges

Which substance alone or in combination and how much would be used, had to be decided based on the thickness of the rock. Knowing Newton's caution, studies may have been completed in the tunnels and galleries. The supporting tunnel piers may have been of less mass than other parts of the works and so require, for example, Vulcan Powder, which had 25 percent less nitroglycerine than dynamite.

Starting September 1, the cartridges began to be made. The drawing on page 124 shows how the cartridge case held, in this example, four cartridges topped by a primer with wires to serve as the igniter. Twenty primers, with fuses and wires properly arranged in a box, with lead and return wires on reels, were carried to each party engaged in this work.

If 49,915 pounds seems like a large amount of explosive, Newton knew "from multiplied experience that it requires from one and a half to two pounds of the explosives … to break up one cubic yard of rock."[12] And more than 63,000 cubic yards needed to be demolished; that is, 170,464,500 pounds of "stratified gneiss … possessing a considerable degree of hardness." Previously, much had been learned in the drilling and explosions at Hallet's Point. The reef had been studied. Thousands of soundings and measurements had been taken. One of them demonstrated that each cubic yard of the Reef would need 0.79 pounds of explosives.

The river itself continued to destroy. One week after the scow explosion, "while attempting to round Hallet's Point the steam tug *J.E. Stevens*, towing a large schooner on her way down the river, smashed into the rocks and stayed firmly fixed to the rocks." In addition, the *Herald* said, the schooner under sail, almost ran the tug down.

Workers on the river may have been assured that the use of nitroglycerine was being re-assessed, no matter how useful the explosive was with its liquid properties. The workers in the tunnels, galleries, and piers had fewer concerns about explosives than those on the river. It is known that the work crew totaled 67 men employed, in this organization: three foremen, 17 miners and chargers, and 47 helpers. One wonders how many of them knew that the roof in the tunnels—the roof being the same as the river bed, was just 13 feet thick. It was then all the experimentation with fuses, with igniters, with cartridges became important.

By the end of March, it could be reported that there had been drilled in the roof 5,375 of the 3-inch holes, and in the piers 1,080 of the 3-inch and 286 of the 2-inch holes, a total of 7,101 holes. The total length of the holes drilled was calculated to be 56,548 feet—more than ten miles—of 3-inch holes and 1,897 feet of 2-inch holes.

Newton's plan, with a few slight changes, principally suggested by General Abbot, Mr. Striedinger, and others, depended on the "vibratory effect" of the upcoming detonation. Why this effect? As the reef was 1,200 feet away from Randall's and Ward's Island and the reef was attached to Astoria, an area then named Long Island City which held a population of around 16,000. The shuddering of the reef at Hallet's point as it exploded might damage nearby structures and kill or hurt those inside those buildings.

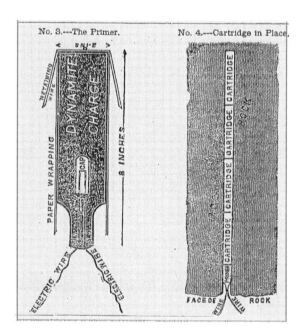

Confused sketch on the left of platinum wire, primer cap, and dynamite cartridges inside their case; but on the right the individual cartridges inside the larger cartridges case is clear (*Frank Leslie's Illustrated Newspaper*, October 7, 1876).

With the complete plan in place, the complexities of carrying out the plan began to be made final. To be addressed were the many tasks to execute the scheme. Awaiting the explosions were 41 radial tunnels, long and short, with 11 transverse galleries, leaving as supports 172 piers of the natural rock to the roof. However, experimentation proved that 80-pound charges of nitroglycerine, fired in numbers of 12 to 20, did not cause a destructive wave, except inside Hallet's Point Reef.

The answer lay in increasing the number of blasts and at the same time increasing the quantity of explosives. This approach, specifically 160 pounds each, settled the problem of reducing the vibrating influence through the reef. Newton could now calculate that "the exterior effect, except an agitation on the water, will be small. So, while the explosive will do their job of destroying the reef, the plan of the detonation will diminish the vibratory and disturbing effect."[13] To take Hallet's Reef apart, Newton reasoned that the primary thought ought to be "to disrupt the piers and roof from their connection with the bed-rock."[14] The piers crumble, the riverbed collapses onto the river floor, and the reef is no more.

## Wiring the Cartridges

Part one of the plan, the noisy and dirty part, was finished by April 1, 1876, and then the drills could be removed as soon as all the holes were made according to specifications.

Once the holes had been drilled in the roof, piers, and walls, there had to be a scheme developed to keep track of the wires that would lead from the explosives to the batteries. First, there would be groups made up of 20 holes. There would be eight groups of 20 holes each (or 160 holes per large cluster). From each of the 160 holes the wires would be directed to just one battery, each of the 23 batteries to be used on the day of the demolition.

Thus 3,680 holes would be connected to explode as one giant blast. In addition, another 747 holes would be filled with explosives which were meant to explode by sympathy. Specifically, 37,000 pounds would be "exploded directly" and the rest by sympathy. That is, the detonation by electricity of the nearby holes (80 percent) would farther explode each of the nearby holes (20 percent) with cartridges not connected to a battery.

To maintain an orderly placing of the charges, Newton and his staff decided that each of the 3,680 charged holes would first have inserted in them wooden plugs, and then be numbered. As an example, we can illustrate one hole's designation and say that on that wooden plug would be written the following numbers and letters to signify the battery, the group, and the individual hole: 17—wire from this hole to go to battery numbered

17 • J—the tenth letter to signify the tenth group in the grouping of 20 •
GG—the seventh letter to show the seventh hole in one group of 20.

So on the plug for one hole might be written 17 J GG and the plug
pushed into the hole awaiting the Rendrock, the Vulcan Powder and the
explosive cartridges to be inserted and this, in 3,680 holes.

Even as the holes began to be drilled early in 1876, the explosives were
packed in tin cartridge cases by the three manufacturers (Rand, Ryan, plus
Varney & Doe) of the Rendrock, Vulcan Powder and dynamite. Newton's
orders for the tin cases included the need for one end of the case to be fit-
ted with a screw-cap with rubber washers so as to prevent the cases from
leaking. Keeping in mind that the job at Hallet's included the charging of
4,427 holes and knowing that General Newton's plan was to use 13,596 cartridges in those holes, some way had to be found to secure all of those tin
cartridges cases.

The solution to secure the cartridges was supplied by Mr. Bernard
Boyle, chief overseer of the work. Boyle saw to it that when the cartridges
arrived, four short lengths of thin brass wire were soldered on the outside
of the bottoms and spread out. The effect of these brass wires was that when
the cartridge was pushed into the drilled hole, the wires, by their elasticity,
pressed against the side of the hole, thereby preventing the cartridges from
falling out, especially those in the roof.

The cartridges were taken down the shaft—perhaps by employing
the derrick, perhaps by a new elevator. Beginning on September 11, 1876,
a "distributing depot" was set up. "Owing to irregularity of delivery," wrote
Newton, "there were at times as much as 15,000 pounds of explosives in
the general depot, and precautions were redoubled until that amount was
reduced within limits."[15]

Twenty primers, with fuses and wires properly arranged in a box, with
lead and return wires on reels, were carried to each party engaged in this
work. This was but one of the safety precautions of the final project since
the explosives were not nitroglycerine alone, but nitroglycerine mastered
by the other materials mixed with it.

"Primer" can be a confusing label and it is true that in the description
of this part the designation is often used to mean "detonator." The primer
also had a fuse. After the holes had been charged with tin canisters of
explosives, the next operation was to insert the 12 ounce priming charges—
dynamite—contained in brass tubes. The primers contained the fuses—20
grains of fulminate of mercury. So now in the holes were likely to be dynamite, Vulcan Powder, Rendrock and fulminate of mercury.

With the explosives in their tin and brass containers now firmly
inserted in each hole, and with fulminate fuse sitting at the open end of
each hole, the remaining 3,680 holes needed to have a mechanism to initiate

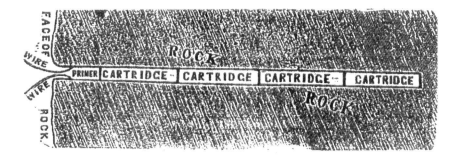

**A very similar sketch again, this time with the rack face in the drawing show how the charges are pushed into the rock itself (*Frank Leslie's Illustrated Newspaper*, October 7, 1876).**

the actual individual explosion. Inserted in each fuse were the terminals of two connecting copper wires coated in gutta percha. Manufactured by S. Bishop and by George M. Mowbray, of North Adams, Massachusetts, the wires would be bundled and connected to the 23 batteries.

Bridging those two wires for each set of cartridges in each of holes was a 0.0015 inch silver platinum wire a quarter inch in length. When a current from the batteries is sent through the wire, electrical resistance causes the wire to heat up. That heat, in turn, heats the platinum. The heat becomes an arc on the fulminate of mercury and the arc creates a shock wave. The shock wave in turn causes the dynamite fuse to explode, sending its force into the cartridges. In this way, the explosion process involving 4,427 holes filled with 13,596 cartridges could all take place in microseconds. On the September 11, the charging of holes was begun, and that work completed at 9 p.m. on the 20th, consuming nine days. Had the cartridges been delivered as agreed upon, this operation would have taken only about four days.

An incomplete image was published early in September and shows (above) how the rock and the cartridge worked together.

To connect the wires, spools were carried out of tunnels and passed by hand up the steep ladder out of the shaft. The lead wire of each of the 23 groups was to be connected with one pole of each of the batteries. The time consumed in placing 3,680 primers, unreeling the lead and return wires, and feeding these out of the shaft, was two days and a fraction.

## Connecting the Wires

Newton's next decision lay in the construction of what was labeled a "bomb-proof building," which seems to have been an accepted synonym for "magazine" or a place for storing explosive material, in some cases bullets. It was pictured in a sketch some little distance from the mouths of the tunnels. Some sources locate it as "three hundred feet from the rock" while others have it as 650 feet.

The bomb-proof building appeared to have been 12 feet square having three main uses. First, it would serve as the place to assemble the 23 batteries, the building into which the wires led. Next, until the time for the explosion was reached, the plates in each of the voltaic batteries were suspended above the battery. Third, the fluid had not yet been added to the battery cases, cases which had cells of zinc and carbon. With the containers placed nearby, the fluid was made in the proportion of six pounds of bichromate of potassa, one gallon concentrated pure English sulphuric acid, and three gallons of water.

To send the electric charge, the batteries were in an arrangement of seven parts, because the vital aspect of the explosion would have to be the synchronized detonation of all of the cartridges. So, the question was how to send the electrical charge to all the platinum wires simultaneously. To ensure this action, Newton and his staff had constructed a device on top of the bomb-proof building, a device connected to the batteries which would send out the charge.

At the lowest part of the device were 23 cups of mercury, the liquid metal that had the property of being a fair conductor of electricity. These cups were contained within a "wooden horizontal disc" to keep the cups stable. The steadiness of the cups was an important feature since something would be dropped into them. As one more safety precaution, the cups would not be filled until just before the explosion.

Above the disc holding the cups was a second wooden horizontal disc holding 23 brass pins, each pin wrapped in wire and connected to a pole of a separate battery. To be sure the pins dropped into the cups at the same time, the disc holding the pins was suspended above, like the plates of the batteries. Holding up the pins' disc was a rope and on top of the rope was an explosive cartridge. Setting off that charge would cause the rope to break, the pins to drop into the cups, the connection to be made between one pole of the battery and the other pole of the battery, thus completing the circuit. All that was needed now was a way to send a signal to the cartridge which secured the rope. One touch of a button would do it.

Newton's more precise calculation showed that once Hallet's Reef was destroyed, the distance from then Hallet's Point to Roosevelt Island would be 865 feet. That may not seem like much room for ships to maneuver, but the space was measured at 638 feet before Newton's project, more than 200 fewer feet.

## *The Danger Remained*

Just as the holes were about to be charged in the piers, tunnels and galleries, a terrible accident occurred. It was the usual practice to transport a limited number of cans of nitroglycerine from Black Tom Island in Communipaw by rowboat, sometimes using a sail, to the works at Hallet's Point. But by 1876 a narrow causeway of more than 1000 feet in length had been built from the New Jersey shore out to the dumping-ground-turned-islet Black Tom, a space whose square footage was wildly estimated. The explosives storage area was situated about nine miles by boat from Astoria.

On the river, the Newton Steam Drilling Scow continued to use nitroglycerine alone. In his typical effort to protect his workers, while using the most efficient blaster, Newton decided to keep the explosives and the scow far separate. Chosen as the safest possible way to convey the explosive, the liquid was cooled, secured in glass or tin containers and then placed in boxes to be rowed or sailed past the Battery and up the East River.

On September 4, 1876, ships passing through Hell Gate south totaled 22 heading north and 23 east. But on September 5, 1876, the *New York Herald* headlined "Blown to Death. A Terrible Explosion of Nitroglycerine at Hell Gate." The paper reported "six quarts of the nitroglycerine—weighing about 15 pounds—was mishandled, dropped, or knocked against an immovable something." The resulting explosion "killed three on that scow and wounded seven ... including a man on a passing tug boat. A changing room for divers, a small cabin, was also destroyed. A rowboat tied alongside demolished. A leg was discovered in a tree."

A witness testified that on September 5 he saw a worker on the scow taking off the top (actually they were corked) from one of the six-quart cans of nitroglycerine just delivered. Just then it seemed that all six cans exploded immediately. "The air was darkened," continued the *Herald*, "with bleeding fragments of human bodies and splinters of timber."[16] The rowboat alongside the scow was demolished and three crew members of a nearby vessel were knocked off their feet. It was reported in addition that a crew member of the "'government tug' nearby was blown into the river and never found." It is enough for the paper to have described one of the victim's injuries as "having the front of his face blown off." As of the issue of September 5, two or more men were thought to be facing death. The scow also sustained serious damage.

The number of workers grew smaller out of fear after this for a brief while. The fear overcame the fact that Superintendent Striedinger said that 100,000 pounds of nitroglycerine had been safely conveyed until that moment, 80,000 pounds of which had been transported just in the past three years. That was at the scow.

Elsewhere, anyone could read earlier that year about two explosions of nitroglycerine. Earlier, at the works at Hoosac Tunnel in North Adams, Massachusetts, two workmen were blown to pieces and three buildings destroyed. One worker died at St. Petersburg, Pennsylvania. General Newton's confidence in "the constraints of the explosion" remained constant as "reporters swarmed over the works" and they could see much of the 220,000 feet of wires snaking through the tunnels and galleries.

One New Yorker feared far more terrible effects. William E. Dodge addressed a long letter to General Newton on his future desecration of the Sabbath. Newton, in what was labeled "a stinging reply" was quoted as saying "I have received a communication from you, dated September 22, in which you decline an alleged invitation from me to witness the explosion at Hell Gate on Sunday, the 24th. As you take a great deal of pains to go out of your way to violate the common courtesies of social intercourse, I take this occasion to inform you that I did not invite you."

A special edition of *The Sunday Mercury* the day before Newton's demolition seemed to be devoted to fear, printing in bold letters "Dynamite Unchained," and asking, "Will It Cause an Earthquake?" It ended its scare leads with the unsubstantiated statement "Scientific Predictions of a Tidal Wave," which then led to the conclusion that there existed "Popular Dread of the Event," dread so powerful that "Yorkville and Astoria Residences Deserted by Their Inhabitants."[17]

## The City Reacts

Certainly, precautions were taken, but the work went on. To be certain of the frugality of his work, Newton again used the large items down in the shaft. Certainly, to be lifted were boilers and debris carts by derrick, and the mules too. It would be typical of Newton to have woodworking and metallurgical tools removed. Surplus dynamite of 5,000 pounds was ordered shipped back to the contractors. With Newton describing that he had scheduled "instructions to invite a certain number of gentlemen," such as the New York Chamber of Commerce, the invitations were put in the hands of Lieutenant Willard, of the United States Engineers.[18]

On September 22, a Friday, Newton went to police headquarters to make final preparations for the safety of all those who might be expected to want to see as close to the explosion as possible. Nearly 1,000 police were assigned to duty on Sunday. It was arranged as well that the police boat *Seneca* would be assigned a few jobs. First, it would carry the mayor, aldermen, and other city officials to Ward's Island, and then act as a guard boat. Police surgeons and their necessaries will be on board. That same day the *Times*

noted that "certain anon-
ymous and alleged engi-
neers" think the event
will be "a much more
serious affair" than New-
ton has been saying. The
article worried that "The
particular of nitroglycerine known as dynamite,
which is to be used in
the. Hallett's Point mine
is one of the wildest and
most untrustworthy
explosives." There were
enough skeptics who
said that the plans could
not work; Newton could
not command power
over so much water and
so much earth and rocks
for the blast to be effective. Persons spoke up
who thought the explo-
sion would shatter win-
dows and they took note
of "the newness of the
explosives as well as the
deaths they had caused."

Drawing of John Newton in the uniform of a
major general, months after the battle of Freder-
icksburg at which he commanded a division. The
illustration is from when Newton was a lieutenant
colonel of engineers in the regular army (*Harper's
Weekly*, October 14, 1876).

Just 30 years before Hallet's Point, nitroglycerine had been formulated, and
it was just 10 years since dynamite had been invented. And Hallet's Point
was connected to the rest of Astoria in Queens.

The *Times* does report that some New Yorkers have fled to New Jersey and Coney Island out of fear. Yet Newton's operations in and on the East
River and elsewhere had relied upon the safe use of dynamite, and the General himself wrote to the *Times* on September 23 assuring New Yorkers of
the safety, stating that "explosives embedded in rock," that is, not in "the
open air," behave differently. Newton tried hard to be clear. "The explosions
of powder mills or nitroglycerine factories in open air" means that the force
of the blast enters the atmosphere, an effect that must "propel waves of air
with great violence." But there at Hallet's Point, where the reef in fact is part
of the neighborhood of Astoria itself, Newton explained, two factors must
be understood.

All the energy, and all of the heat and all of the gases will be communicated to the rock and the water … and … the mine when flooded will hold 47,461 cubic yards of water—10,251,576 gallons of water—and the volume of water above the reef at high tide is estimated to be more than twice as much. The explosive force then must move 63,000 cubic yards of rock [170,100,000 pounds] and a dead weight of 143,000 cubic yards of water [386,100,000 pounds].[19]

Thus, the river itself will have a tamping effect, a lessening, on the release of energy into the air. All of the energy released by the blast will work only on the rock and the water. Nevertheless, some residents feared the shock of the explosion; some feared the release of poisonous gases. Part of the dread came from the fact that the work that had been done had been out of sight, all the way down the shaft and under the river itself. Newton attempted to allay the biggest apprehensions by noting "there will be no commotion of the air and the underground effect will only be sufficient to jar buildings."[20] For those New Yorkers living near the river,

the lighter articles of furniture were taken to the piazzas and the heavier placed in the centre of the rooms and covered with mattresses…. In one solid stone house … beds and … all the other furniture, even the carpets and door mats [were taken] to the back yard … a cook stove [was] deposited beneath a large apple tree."[21]

Even though many people left the city, one positive aspect lay in the prediction of plate glass installers. The superintendent of a gas house on Manhattan Island visited Newton and was assured, as were the brewers in Harlem Heights, that "the shock of the explosion from 200 feet away was likened to the quaking on a street when a truck rumbled through." Newton found it necessary to assure everyone that the cartridges in their holes "lie as comfortably there as babies in their cradles. Cool and damp as they are, you could not explode them if you tried."[22]

# TEN

# Newton's Triumph of Science

## *The Crowds Assemble*

The rain that day did not stop New Yorkers from gathering in the rain, waiting for the big moment. It may be that some of them had read a headline "Hell Gate Destroyed in the Explosion."

> Most of the talkers had implicit faith in Gen. Newton, but there were croakers and doubters, who caused some uneasiness, expressing forebodings. The saloons did a big business. The churches were very poorly attended, but that might partly have been owing to the rain. The Rev. Dr. Harris announced from the pulpit that, for obvious reasons, there would be no Sunday-school or afternoon service.[1]

Notwithstanding Newton's calming statements, on September 23, the day before the blast, "Great alarm prevailed amongst residents of First Avenue" in the blocks opposite Hallet's reef. "And some houses were entirely deserted … through a fear of rocks being hurled across the river or a huge wave being caused by the upheaval."[2] Those who had not fled opened their doors and windows, fearful of the power of an air blast.

That Saturday night, workers assembled to finish the preparatory work. With the wiring to the batteries checked about midnight, part of the coffer dam was opened. This flooding through a small opening using a siphon pipe was made to flood completely the tunnels and galleries. It was completed as a part of a two part process "in order to increase the energy of the explosion by somewhat closing the vent [opening] of the holes, or tamping [to concentrate the force of the explosion] by the pressure of the water" by 2 a.m.[3] (At that same hour, the wires were connected to the batteries.)

Six watchmen, armed with ax handles, stood guard at

> the government works as the big day dawned. This was the day that General John Newton had arrived at through his thoroughness in keeping costs down, in always looking for, and buying, the newest and most efficient tools, and by insisting on a level of safety that ensured the success of the project.

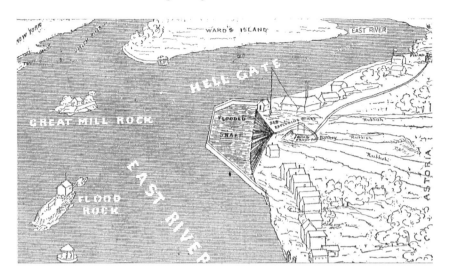

**Even as demolition day neared, this map shows a shack on Flood Rock. Newton's next project, since some preparatory work had begun on the tiny piece of the nine acre rock that could be seen above water.**

He concentrated on that day on the work to be done and nothing else, noting while the staff that Sunday morning saw to it that everything was ready for the detonation.

General Newton had ordered the installation of "two insulated copper wires leading from the battery in the bomb-proof to the hastily erected platform on the northeast shore of Pot Cove [the 'firing point'] from which the signal to the batteries was to be fired, … carried through trees and over fences, following the trend of the shore." The platform, about 15 feet high, was erected about 700 feet from the Reef at "Mr. Ramsey's place on Pot Cove [address unknown]," 2,200 feet from the explosives. On Saturday, the *New York Sun* wrote, "the opinion was freely expressed that the bomb-proof was not a safe place to be in, and some persons believed it must be blown up and the batteries with it when the mine was fired."[4] Troops with fixed bayonets were stationed around it. This small building was but one more intermediary between the firing point and the explosives.

Above the bomb-proof building was the device of 23 keys dropping into 23 cups of mercury suspended from a rope. Those troops stationed nearby might be able to see the explosive cartridge atop the rope. The signal to explode the cartridge would be sent from the firing place to the bomb-proof.

Two of the 60 workers on the project were selected to finish the connecting of the electrical circuits of the 23 batteries. All 60 had volunteered to do the job. The plates for the battery had to be at hand, cleaned, and ready to insert. These men, some of whom have worked for a decade on this project, had completed their task.

The tugboat, *The Star*, was made ready at 6:30 a.m. Saturday for any possible assistance it might provide. With the tunnels and galleries filled with water by 7 a.m. Sunday, the remaining preparations began at 8:00 in the morning. Many hands remained busy, though rain poured down hard most of the time. The fluid in demijohns for the batteries—150 gallons of it—were made ready to pour. The crew with batteries and wires approached the bomb-proof building, 12 feet square, in which Newton had ordered the batteries to be placed. The metallic plates were not submerged in the acids till noon so that the electricity could not be generated till near the time of the explosion, which had been fixed at 2:50 p.m.

Though it rained steadily that day, some people left their houses and stood out in the weather in the middle of the street rather than, as they feared, being trapped in the house as it was blown apart by the force of the explosion. Guard boats began to position themselves at four points on the river to alert and halt East River traffic, and the police took up a line along the Astoria riverfront to guard against a "too near encroachment" by sight-seers.

People continued to fill the nearby rooftops of the tallest buildings, including a few churches. By 8 a.m., the roof of the tall George Ehret's 1871 brewery building (between Manhattan's 92nd and 93rd streets and Second and Third avenues) filled, placing them seven stories above the river. Passengers on an elevated line (probably the IRT Third Avenue Line) might be able to be in sight of the event.

By the next hour, wiser viewers had climbed aboard Third Avenue horse cars in Manhattan. Thousands went to Central Park to be far from danger and still others went to streets in Astoria looking for the closest viewpoint. Soon 100,000 to 150,000 people had gathered at various spots. City officials and police were stationed in case of a panic among the viewers. Meanwhile, the General's wife, brother, daughter, and son were taken to the scow anchored off of Ward's Island by the steam tug *Juaniata*. Those watching the workers could see that "the same order and method prevailed in the closing part of the work that had characterized the whole enterprise from the moment the government took possession of Hallet's Reef. There was no confused hurry, but everything was done is a quiet, systematic manner and yet with rapidity."[5]

Probably the best viewing point in Manhattan seemed to be from Newton's steam drilling scow, where many officers of the Corps of Engineers and their ladies had been taken to witness the explosion. Some of the civilians in Queens selected Hoyt's Place, about 700 yards east of Hallet's

Point because the spot there is high, the river only seemed to be 200 feet away. No matter how heavy or light the continual rain, "fully 200,000" have eyes fixed upon the reef.[6]

General Newton chose to arrive at 11, and while he was absent "Mr. Striedinger, the electrician, never losing his temper or speaking an impatient word, directed every step … with a cool and collected precision."[7]

From Manhattan, at 11:30, Newton crossed to Astoria on the 92nd Street ferry. He told workmen that he passed a quiet night and never felt better. The *New York Sun* wrote,

> … he is a soldierly-looking man, with hair and beard slightly sprinkled with gray, and firm Roman nose. He appeared to be very cheerful and showed no trace of nervousness or anxiety. "I've been five years working here, getting ready for that one supreme instant," he said, nervously clipping his words, "I've seen 52,000 pounds of explosives put in, and oh, now, I'm waiting for that one minute to see how she'll break up."[8]

Newton had seen to it that there were two small steamboats—a small launch and a heavy tug—to ferry men and material as needed. For now, his wife and daughter remained on the scow.

Preparations to take depth soundings began after the explosion were finished. Near mid-day, at the "Battery House" (another name for the "bomb-proof"), the demijohns holding the battery acid were emptied into the 23 batteries' cells. Then the metallic battery plates were submerged in the acids so that the electricity could be generated. The blast area was closest to Pot Cove.

One telegraphed account sent all over the world noted "'the immense crowds from all over the city' to view the explosion." There were "thousands and thousands of carriages" for those "at ease with the danger that still others foresaw. Among many residents of First Avenue," for 15 blocks up to 96th Street, there remained an uncertainty about the effects of the blast and not simply on their glass window panes. Many were fearful of having their homes attacked by "rocks being hurled across the river" from Astoria or having "their homes and selves being swept away by 'a huge wave.'"[9]

The size of the wave did not seem to scare the "nearly 100 small row boats [which] lay along the Manhattan shore above and below the guard boats." Others not quite so brave "feared the effect of noxious gasses that were to be liberated in hundreds of thousands of cubic feet."[10]

Horse cars traveling along First and Second avenues, and the housetops along the East Side were crowded. "Thousands of vehicles of every description filled the heights and cross streets." Continuing to set the scene, a telegraphed report took note that "every house in the vicinity had its doors and windows open, as a matter of precaution, but the great mass of people had no fear whatever, and spectators swarmed on the piers and low grounds directly in line with the rock, where they would be drowned by thousands in case of a wave."[11]

Police lined First and Second avenues and were stationed among the crowds. Between Manhattan and Ward's Island, opposite about 11th Street, government scows and the police and Emigration Board steamers could be seen along with hundreds of barges and rowboats. Guard boats were stationed on the river to "force a suspension of navigation." Four guarding steamboats were placed between Blackwell's Island and Astoria, and they also served the purpose of accommodating 800 invited guests.

About 12 o'clock, streams of pedestrians crossed the city to the East Side; vast numbers of strangers came to see what everybody "expected would be a magnificent spectacle," including those on board the many vessels nearby, such as the *Neversink* and the *Flora*.[12]

By 1:30, the different guard steamers had arrived, as had other steamers with officials, commissioners, and others, and as had the last excursion steamers from the Morrisania and Harlem lines filled with spectators. Some small steamers and tugs with names like *Sylvan Dell*, *Harlem* and *Sunshine* were packed with curious passengers. The Society of Civil Engineers chartered its own steamer, the *Pleasant Valley*, and took up position near the blast. Representatives from both the British army and the French Academy of Sciences were watching as were civil engineers from Russia, Germany, Austria, Brazil and elsewhere. There to observe were four priests (Newton was a Roman Catholic convert), senators, and a Supreme Court judge. "Finally, the hour for the blast drew near, and the clicking noise of opening and closing watches was kept up by impatient spectators."[13] Watching from many spots were reporters from at least five newspapers: *The Herald*, *The Sun*, *The Brooklyn Daily Eagle*, *The New York Times*, and *The New York World*. Stories would go out all over the world via the telegraph. Yet, including scientists,

> … none were free from a haunting anxiety concerning the issue of the experiment.… On board was … George M. Mowbray, the North Adams nitroglycerine manufacturer and early experimenter with the explosive at the Hoosac Tunnel.… Mowbray said that "Gen. Newton … required not only a lifting, but also a rending, splintering force" and so had chosen three different formulations with nitroglycerine.… Mowbray added, "The passage of the igniting spark will be so rapid that the space of time that it occupies in passing from group to group will be imperceptible."[14]

By 1:30 all the four guardian steamers, ships to warn off any maritime traffic, had moved to their positions. And by that time, "two millions of people in New York, Williamsburg, Westchester, and Jersey City are listening almost painfully for the dreaded thunder roar."[15] Those millions could only guess, could only imagine what might happen as 50,000 pounds of dynamite were ignited. Certainly, the 200,000 onlookers standing in the rain within sight of Hallet's Reef would have been the most apprehensive. Reporters just from the *New York Herald* were sent to file stories from 94th Street, Ward's Island, Governor's Island and the Croton Reservoir (later the site of the New York Public Library). Spectators could be found on steamers not far from Hallet's, from the foremast of the ship *Savoir Faire* at the foot of Grand Street, Hunter's Point, Williamsburg, and from *Belvedere Tower* in Central Park. In all, an "immense concourse of people present."[16]

Ward's Island was considered a favorable place to view the explosion, and the applications to go there "were very numerous," but Newton's invitations were limited to a select party. On 90th to 109th Street on Manhattan's East Side the grassy area on the river was black with spectators, as they huddled under their umbrellas in the September rain.

General Newton's launch *Flash* glided back and forth as needed from the reef, to the scow where three soldiers were manning a cannon which would serve as a warning device alerting all to the imminent explosion. Also, a man with a white flag had been positioned on the scow to signal if all is clear on the river. Now the 23 cups were filled with mercury awaiting the contact with the brass pins. Mr. Striedinger checked on the "two insulated copper wires leading from the battery in the bomb-proof to the hastily erected platform on the northeast shore of Pot Cove, from which the battery was to be fired."[17]

New Yorkers had been informed there would be three firings of a cannon, the last one as the final notice of the impending explosion. "The crowd seemed oblivious to the pouring rain," the *Herald* wrote, "so eager were they to see the large blast." On Ward's Island, there was some talk "of what many regarded as an exceedingly doubtful experiment."[18] In Manhattan, along Third Avenue, seemingly all nearby housetops were full of watchers, and on Fourth Avenue, with its relative height, were gathered "a great throng of carriages."

At 2:25 p. m. the first signal gun was fired from the government scow. As the 2:30 time was viewed on watches, some people, too far from a good viewing point, "started on a brisk run for the river."[19] Arriving at the firing point, about 650 feet from the mine, then were generals Abbot and Humphreys and they joined Mr. Striedinger, Captain Mercur, Lieutenant Willard and supervisor Boyle, their duties completed. At 2:35 a second warning shot announced the near approach of "the dreaded hour." "'Thousands of vehicles of every description crowded the heights and cross streets' as the blast drew near" as the rain came down in torrents.[20]

## *Blast Time*

At 2:40, after about a three hour inspection tour, Newton and his staff traveled back across the river to the scow. There Newton saw to it that his wife Anne, a nurse, and Mary Anne Newton, the general's not yet three-year-old daughter, were safely aboard a tug for the trip back across the river to Astoria. When they arrived at the platform's firing point, all was ready. One newspaper citing "the following interesting incidents" from other papers, described "General Newton ... arriving at the firing station led his wife up the pier and the nurse followed with the little girl whose fairy fingers" were to set off the explosion.[21] "Mrs. Newton was given a small stool on which to sit. Some officers and civilians stood outside the firing point shed but close at hand, while General Newton frequently consulted his watch as he talked in a pleasant way with Mrs. Newton and his staff."[22] "Gen. Newton frequently consulted his watch as he talked in a pleasant way with Mrs. Newton and his staff."[23]

All was ready now. The *Herald* thought "everyone except the general [Newton] looks anxious and expectant."[24] The drilling of the holes had been completed six months earlier. The wires from the drowned explosives and their platinum wires exited the tunnels and been fed to the poles of the batteries in the bomb proof on September 20. The batteries had been filled and the plates inserted at noon that Sunday. The five-part mechanism atop the bomb-proof had been checked: the mercury-filled cups awaited the descending brass keys; the rope suspending the keys hung in place, and the explosive holding the rope had been checked. All the safety precautions, save one, were removed. What remained was the wire to the platform firing point that would set off the explosive and drop the keys. That wire was made ready at the platform.

Then the presence of the young child began to be noticed by the crowd, she seeming to be so out of place. The "'baby seemed to inspire awe and wonder among the gazers.' Lieutenant Willard whispered to the man at his side 'The baby is going to fire the mine!' The listener wheeled instantly toward his next neighbor and whispered in turn, and that electric message flashing through the crowd exploded the mine of curiosity in a burst of surprise."[25] It was Newton's surprise to all that the baby would press a telegraph key to close the circuit and set off the massive explosion.

The third cannon fired. For a short time, Newton thought the ship *Pleasant Valley* was going to drift right next to the explosion but then it halted. Captain Mercur with binoculars saw a white flag waved from the scow anchored off of Ward's Island and the officer shouted "Now." Lieutenant Willard, running out upon the pier, waved a handkerchief for the last signal to Colonel Abbot.

Inside the platform at Hoyt's a telegraph key, a beautiful object, had been crafted from rosewood, and on the rosewood has been put a plate that read "Lieutenant-Colonel John Newton. United States Engineer Corps. Hell Gate September 24, 1876." The key itself rested on top of two less elegant empty dynamite boxes. The telegraph key, was ready to fire off 3,680 explosive cartridges, holding almost 50,000 pounds of explosive.

Newton, precise as usual, consulted his watch as he watched the seconds tick away and then said, "'Come, my little girl.'" So that the two-year-old Mary could reach the key, she was placed on one empty cartridge box. She was being watched by Mr. Striedinger, who had worked on this project for so long. Newton now had in his left hand his pocket watch and in his right his daughter's tiny hand.

"'Come, what are you going to do? You have been talking about it for some time.' As Mary's hand was then allowed to approach the key, Newton said, 'Come, come, look out there toward the blast. You don't want to look here.'"[26]

"As the little girl's hand now touched the telegraph key, all eyes, feeling a tremor, next could hear a muted rumbling sound and at two hours fifty minutes PM, September 24, 1876 at high tide the water above the reef began to roll as if in a boil" And "the little noise was a surprise to everyone."[27] Some thought the noise might be called a thud or a rumble.

Though there was a lack of agreement in the maximum height of the explosion of Hallet's Reef, most reports guessed the eruption at a height of at least 75 feet and was comprised of foaming water. The *Herald*'s man reported that he saw "a dense column of black water with a horrid crest of shivered beams and splintered rock."[28] The second stage seemed to be a mix of black or yellow smoke, mud, and large pieces of the coffer dam.

Another source was more specific, writing, "The commotion of the water was great for a few minutes, and the earth of which the coffer dam was partially composed made it very muddy, while pieces of timber and boards were thrown over and over on its surface in such a manner as to show how greatly the element was stirred."[29]

## Reaction and Judgment

Most newspapers wrote that the viewers on the shore and on the vessels uttered not a word, hardly breathing. "They stood in perfect silence watching the spot where more than 50,000 pounds of powerful explosive had been fired." But very soon "[t]he cheers that broke from the crowds were deafening, while the whistle of every steamer was employed to increase the sounds of rejoicing. 'Three cheers for Gen. Newton,' someone exclaimed,

**Sketch picturing young Mary Newton and her "baby fingers that exploded the mine." Mary uses a telegraphic key to close the electrical circuit while her father and Mrs. Newton stand behind (*London News*, October 1876).**

and they were given and prolonged by loud and enthusiastic shouts. A mighty cheer arose from the northeast side of the city"[30] as well as from Astoria, said a syndicated report, "and from steamers and boats, steam whistles were blown and bells were rung out joyously."[31] The *Herald* simply headlined "A Triumph of Science."

One post-blast report had the general "smiling" and then "turns to the party on the platform and says, 'That's something like an explosion, gentleman,' the officer clearly pleased with his work." The feared wave did not appear, and the disruption of the river was "no more than it would have been by the passage of a large steamer."[32] Newton's science had confined the 25 tons of explosives to the rock and the river while keeping the explosive force out of the air where it might cause some damage.

Another reporter stated that "in restricting the effect so closely to the immediate neighborhood of the reef, though it cannot be forgotten this was an explosion of fifty thousand pounds of powerful explosive, General Newton has shown a control over these hitherto untamed explosives which marks a new era in the history of such matters."[33]

"Yesterday's triumph leaves the Engineer [Newton] still standing in the front of human effort and achievement," wrote the *Herald* on September 25. "The magnificent spectacle," it continued, was "a triumph of science."[34]

"Hundreds of row boats quickly set out over the blast site for a profit might be made from Newton's achievement. Dead fish were scooped up, as were anything connected with the explosion: bits of rock ... pieces of wire, pieces of the dam." Many "were taken up and cut into small pieces for mementos [sic], many of the pieces being afterward sold for twenty-five and fifty cents an inch."

*The New York Times* noticed that "three or four row-boats were pulled round [Hallet's] Point while a very heavy ebb tide was running ... at about seven miles per hour. The men about the neighborhood declare that this had never been done before. And yet three quite young lads managed ... to succeed in pulling round."[35]

Then the General and his staff rode out in the launch from Pot Cove and around the bend to Hallet's Point to look at the results of the explosion. We can understand his pride in the work begun more than ten years before but knowing Newton's professionalism we also know that he wanted a close look at the remains of the blast.

He had ordered the steam-tug *Clara* to be prepared to take soundings. At the same time he saw to it that his orders were followed quickly for buoys to be placed to delineate the ruins of the reef. Those markers would help to outline where the dredgers had to begin their work. The depth of 26 feet would not be reached until the debris from the blast had been scooped up and taken away.

Caution was impressed upon the seafarers, since broken rock occupies 60 percent more space than solid rock, but the eddies were lessened once Hallet's had been demolished. "The figures show that four hours after high water the current at Hallet's is four times greater that of Throgs Neck at the opening to the Sound."[36]

We can rely on John Newton's report for the more scientific description of the explosion: "The elevation of spray, vapor, and gases projected upward, reached to the height of 123 feet.... The explosive effort in the air was not perceptible.... Along the line of the reef a little plastering was dislodged from a ceiling in a house 150 yards, and in two houses 600 yards, from the work."[37]

As far as *The Commercial Advertiser* was concerned the detonation was labeled "A Complete Success Two Hundred Feet of Shore Line Swept Away."[38] A day later the *New York Tribune* concluded that "The work proves that recently discovered natural forces—forces far transcending in power any that were before known—may be made to obey the human will."[39] Every cartridge had been exploded. The major conclusions were two:

1st. That an unlimited amount of explosives distributed in blast-holes in moderate charges, proportioned to the work to be done, thoroughly confined in the rock and tamped with water, may be fired without damage to surrounding objects. 2d. That an unlimited number of mines may be simultaneously fired by passing electric currents through the platinum-wire bridges of detonators.[40]

This explosion was not simply getting rid of a troublesome reef. Newton's engineering feat had many implications in other places and other times. The day after, in one of its headlines, the *New York Herald* wondered "what commerce will derive from an engineer's achievement. Part of the shipping will no longer have to face such a tight bend in the river, as the explosion will now broaden the river to 600 feet at Hallet's."[41] There remained those who stayed cautious about the demolition. "But if the improvements were made simply for the benefit of the coastwise trade, no doubt can be entertained of their importance and utility."[42]

There were no doubts either about the explosion or the extended consequences of Newton's work.

> This vast work was accomplished too without loss of life or the slightest damage to property, and the croakers who predicted an awful disaster were effectually silenced. The success of this experiment, for such it was, will open a new era in engineering and the same method will come into extensive use in deepening channels and removing obstructions.[43]

One journal took a broad view. *The Internal Revenue Record and Customs Journal* thought of the less impeded flow of water down the river. "Gen. Newton's work has increased th[e] difference" in the speed of flow all the way into the harbor and past the narrows, even "for an improvement of the Sandy Hook entrance," 16 miles away.[44]

This explosion affected much more than New York because the interest in the project appeared in papers all over America and overseas as well. The event was printed by *The New Hampshire Patriot*, *The Duluth Weekly Tribune*, *The San Francisco Daily Evening Bulletin*, *The Prairie Farmer*, *Forest and Stream*, *Scientific American*, *Maine Farmer*, and *The Illustrated London News*.

As a little time passed, commentary about Newton's achievement covered a narrow observation about the East River exploit as well as ideas about what the success of the explosion might mean in the future. Most writers explicitly or implicitly credited Newton if for no other reason than New Yorkers had pleaded for the destruction of Hallet's reef for at least forty years. Some writing may seem exaggerated, as was the style of the times.

What was not exaggerated as perceived by a ship's captain was that "the decrease in the swiftness of the tide in consequence of having a broader channel is greatly in favor of the safety of sailing vessels going through The Gate on either side."[45] The 92nd Street Ferry found it easier to negotiate the slips on both sides of the river and the eddy off Flood Rock had shrunk by 25 percent.

The writer Thomas Wallace Knox wrote in 1877, "Newton may be said to have achieved a victory greater than any in war. The 24th of September ought hereafter to be celebrated as the jubilee day of commerce, and the anniversary of the greatest triumph of engineering science in history. The seven years' toil on Hallett's Point Reef at Hell-Gate has culminated in a grand coup de maitre," or stroke of a master.[46] Reflecting back almost a decade later, *Farrow's Military Encyclopedia* looked "upon the stupendous operations of Hell Gate with wonder and admiration."[47] For *The Internal Revenue Record and Customs Journal*, "the seven years' work at Hell Gate was valuable to every maritime nation. It is not impossible that such circumstances will arise as may make it the first in a chain of engineering operations that may alter the highways of commerce."[48]

Two months after the explosion *Popular Science* thought Newton's work "must be regarded as the most brilliant piece of scientific engineering that has yet been accomplished." And *The New York Times* front page labeled it "another triumph of human skill over the resistance of nature."[49]

The East River Bridge—The Brooklyn Bridge—was halfway through completion.

The *Times* not only praised the work but even suggested that the river be renamed "Newton's Channel,"[50] and it was not the only New York entity to praise the work. One letter suggested contributions to "a memorial" to Newton. The Chamber of Commerce looked upon the scope of the work on the river resolved "recognising in the removal of Coenties and Diamond Reefs, in the harbor of New-York, and of Hallett's Reef at Hell Gate, with what skill and efficiency, and with what safety dynamite and other powerful agents have been employed, tenders to General Newton and the engineers associated with him its congratulations and thanks."[51] The New York City Produce Exchange and the Board of Aldermen also formally offered congratulations.

For Newton, the greatest honor came on October 23, 1876, when the National Association of Science brought him into their midst as a full member. "The membership only came about through election by his peers to an organization 'committed to furthering science in America, and its members are active contributors to the international scientific community.'"[52] General Newton may be the only NAS member elected for achievements in destruction, though it's not possible to say that with certainty.

With work still to be done to measure the effect of the blast, William Preuss, the surveyor soon made 120 soundings. The remainder of the years, that is, from October 7, 1876, to year's end included the spending of $20,606.61 for drilling a few holes and the cost of a steamboat, material for building a seven foot high sea wall, and building a dock 40 feet long.

Thus, by year's end, from the material excavated by the workers, was built "Flood island on which they made the base of their operations. This island is about half an acre in extent, and is, or has been until within a few days, covered with buildings containing engines, steam-pumps, hoisting apparatus, machine-shops, and all the appurtenances of a complete mining plant."[53]

# ELEVEN

# Newton's Aim of 26 Feet

## Newton's Second Decade

The year 1877 marked Newton's attention turning mostly to the demolition of Flood Rock even though June 7, 1875, was the date when the sinking of the shaft was commenced. Flood Rock, nine acres of gneiss, measured 1,200 feet long by 602 feet wide, but only 250 square feet width of that reef could be seen above water. Seen, that is, if the weather presented no fog, no snow, no rain, etc. If seen, could the pilot successfully avoid the destruction of the vessel and its cargo. We know how frequently the pilot did not. One out of every 50 did not.

The General could not yet know it but by the time the demolition would be finished, and that would not be until 1885, almost a billion pounds of rock would have to be demolished. In mid–East River, at about 1,000 feet northwest of Hallet's, Flood Rock loomed about 1,600 feet from Manhattan. "The summit of Flood Rock … an area of about a quarter of an acre was being prepared for the necessary buildings and a hoisting tower at the opening of the shaft," reported one scientific journal.[1] The debris was first to be taken by dump scows to a deep hole off 92nd Street, so that it might be filled to a desirable level.

Much of the work in the river had taken place in what was sometimes called the Middle Ground, that part of the river northeast of Hallet's Point in mid river that surrounds a line drawn from mid Pot Cove to Scylla Point on Randall's Island. However, Newton could cross off the work that had been mostly completed in and around Hell Gate, some 16 obstacles. At this point, these obstacles up and down the river can be tallied below.

*"Middle Ground"*—Shelldrake • Pot Rock • Scaly Rock • Way's Reef • Frying Pan
*Eastern Channel close to Astoria*—Bald-Headed Billy • Bread and Cheese
*Near to Ward's Island, north of Hell Gate*—Hogs Back and Holmes Rock • Heel Tap Rock

*South of Hell Gate close to the Battery*—Diamond Reef • Coenties Reef
*Fenders installed on the Middle Reef*—Hen and Chickens • The two Negro
   Heads • The Gridiron

   In addition to those 16 and Hallet's Point Reef, the Newton scow had
been set to work on Ferry Reef, also in the East River at 34th Street; Baxters
Ledge in Kill Van Kull near Newark Bay; Corning Rock, in New Rochelle
Harbor, and upon a channel through a reef in the Harlem River.

## Some Results

   Meanwhile, following the explosion at Hallet's, early examinations in
the fall of 1876 and early winter of 1877 included the use of sounding apparatus to check the thoroughness of the 1876 blast. Those measurements
showed a depth of from 18 to 20 feet at low-water. Dredging, which process
might take 18 months, was needed to reach the required depth.

   The steamer *Providence* of the Fall River Line which could now sail
about 50 feet closer to Hallet's Point than any steamer ever before. The additional width at the opening to the Eastern Channel—200 feet more—would
keep crafts from foundering on Flood Rock. It appeared that the blasting
had produced channel space.

   The debris of the 63,135 cubic yards of solid rock shattered by the Hallet's Point explosion now measured 87,000 cubic yards. General Newton
estimated it would be safer and more economical to remove the debris,
down to the depth of 26 feet, by grapple dredgers. As only blocks of five
tons could be easily raised, and blocks exceeding ten tons could only be
lifted under favorable conditions, a good deal of the rock had to be first
reduced in size by additional blasting, increasing the cost. But like debris,
the exploded debris from other rocks and reefs still rested above the waterline. That dredging work would continue for some years as well.

   Two years earlier, 1875, the Flood Rock project had begun on June 9 by
the first operations of sinking two shafts 65 feet apart, each being carried
down 60 feet below mean low water. The larger of the two measured 10 by
20 feet and the smaller shaft 12 feet square. Each shaft would serve a different purpose. With the construction of stairs still two years away, the miners
were lowered using the crane that stood at the shaft mouth.

   Meanwhile, a complete consensus was not reached on the expense of
the Hell Gate projects. For example, 1886's *Appletons' Annual Cyclopaedia
and Register of Important Events* suggested that

   The fogs of Long Island Sound, with its crooked channel and rocky shores,
   would prevent the ocean-ships from using this road for the sake of the 200 miles

that would be gained; and the risk and inconvenience of crowding the East River with shipping would be another deterring cause…. But if the improvements were made simply for the benefit of the coastwise trade, no doubt can be entertained of their importance and utility.[2]

## A Challenge

While the opinion of *Appleton's Cyclopaedia* might have been ignored by Newton and his superiors, a louder voice came from Commodore Vanderbilt, then 80 years old. Successful in both shipping and railroads, the rich man had opinions on many topics that attracted attention and it was shortly after the Hallet's Point detonation that the magnate loudly doubted that the East River would serve as the major passageway for ocean steamers. Only the Sandy Hook and Narrows route would be suitable, he said. By ignoring the traffic to New England, to Canada, and out to the Atlantic via the East River, Vanderbilt left out many companies, many ships, many passengers and much freight.

Soon Vanderbilt found that the *New York Herald*, in a single article, pointed out the many advantages in speed, safety, and comfort afforded by the Sound/East River route.[3] First was the 50 miles of distance saved from sailing to The Narrows entrance and second was the 150 miles of smoother sailing waters on the Sound rather than the Atlantic. Then too the "coasters," steamers from three companies like the Fall River Line, sold luxurious accommodations connecting travelers first from New York City to Connecticut, where connections were made to Boston, Worcester and Providence by rail.

And, should a traveler journey from England to Boston, how to get to New York City? That meant the time and expense of waiting at Sandy Hook for the tide. Vessels there crowded at anchor together must wait because there was no truly safe and speedy passage into the Port of New York from the Atlantic until 1907.

## Newton's Argument

The arguments may have gone on but Newton's fiscal year's appropriations of $250,000 were halted on December 31, 1876, and did not begin again until the beginning (i.e., July 1) of the fiscal year 1878–1879. Newton wrote

No appropriation having been made … it became necessary to suspend all work at Flood Rock, except such as was absolutely necessary for the preservation and storage of materials. The only work ... consisted in overhauling and storing the machinery, rebuilding a portion of the sea-wall torn down by ice, moving the laboratory and water-tanks, whose foundations were washing away owing to insufficient protection.... As soon as this work was accomplished, the employees were discharged and the works left under charge of a watchman."[4]

Although mining could not go on, upkeep, replacement, and installation of machinery went forward, since the new equipment was paid for. So, the works on Flood Rock saw installed a new surface condenser, a planer and refurbished compressors, a lathe, and a drill press. In addition, a contract was signed to take "rubbish rocks" to the space between the rocks called Little Mill Rock and Great Mill Rock. This dumping would also erase the eddies flowing between the two. Over time, the tonnage of rocks became so great, that in 2020 on a Google Map you can see the two Mill Rocks as one Mill Rock between Rhinelander's Reef at 91st Street in Manhattan and Heel Tap Rock, a rock closer to Ward's Island.

In the spring, the General was called upon, as he had been called upon frequently, to discuss his work and offer advice at the request of many groups. Some of those assemblages included the Federal Light-House Board, the Board for Determining Positions for Atlantic and Gulf Interior Coast-line Defenses, the House Committee on Rivers and Harbors, and the New York Department of Docks.

And since he worked and lived in New York City, he agreed to speak at the Cooper Union on 7th Street in the East Village on April 11, 1877. Prepared to address the 900 who attended, he illustrated his speech with drawings and "numerous apparatus." His talk included his favoring of improvements to the Harlem River.

But the central purpose of the talk was to explain his shattering of Hallet's Point in the fall of 1876. Newton explained that the most efficient method of demolishing Hallet's Point would be to take all of it down at once. But to do that, some way had to be found to detonate the thousands of "mines" (that is, "explosives") at once. He acknowledged the contributions of General Abbott for his experiments with electromotive force, the resistance of the battery, fuses and wires and others at Willet's Point. He also showed his audience a small demonstration of how the Hallet's Point explosion worked, using a plaster of Paris miniature model of one of the tunnels.

## Delays at Hell Gate

Newton remained aware that his small fleet of tugs, scows, and smaller ships has had to deal with dangerous things. Natural phenomena presented peril for his little fleet just as the commercial ships did: fog, ice, wind, snow, currents, tides, eddies, and whirlpools. Some in the city, unaware of the budgetary restraints and the complexities of his task, might complain of the slowness of the work on the East River. Newton was quick to point out the reasons for suspensions in Hell Gate:

First, the obstacles remained, obstacles that themselves caused the problems like eddies and whirlpools that could only be eradicated by destroying those very obstacles. The largest one, Flood Rock, remained. The river was characterized by a swift tide, submerged rocks, narrow channels, and cross currents of 6–8 MPH.

Next, he had begun his work with very inefficient means such as drilling by sledgehammer.

Third, until there were some improvements in diving and the invention of his scow, surface blastings did not accomplish much. Even then, it took some years for the Newton steam drilling scow to be developed, accepted and used.

Fourth, it was not until 1866 that the government was willing to pay for a thorough examination of the river and to determine the cost for it. And the money came in small amounts.

Lastly, the intricate formula whereby all the electrical wiring and signals would act as one took years to be finished through the labors of Newton, General Abbot, and Mr. Striedinger.

To add to all of the problematic items, the year 1877 was more than four years into the Long Depression that began in 1873. It would last a year and a half more. Newton could not know how long the Flood Rock project would take to complete without funding, but with no money available until July 1, 1878, a year and a half had to pass.

This hiatus in funding allowed Newton a respite from his river work and so he was listed as on a leave of absence in Europe, April 21 to August 20, 1877. Temporarily in charge was his assistant Capt. J. Mercur, left with little cash on hand, and many of the problems still in place. "Although this Hallet's reef was blown into history, some few perils, in the shape of rocks and reefs, still remained; and to these the attention of the engineers was next directed," *Harper's Weekly* pointed out.[5] One problem was the most comprehensive to deal with The Gridiron—to at least make it more visible—so "The rock removed was dumped in the shoal-water upon Flood Rock and the Gridiron, for the double purpose of covering the Gridiron, as an improvement to navigation," Once more The Gridiron and Flood Rock seem joined.

The river, of course, remained dangerous. On May 2, 1877, the 72 ton schooner *Jennie C. Russ* struck a Hell Gate reef and sank in 15 minutes. The ship and her cargo of coal were a total loss.

## The New York Central

Since 1853, one man had been building his own way from New York City to Albany and across the Hudson and then West and East. By buying a series of separate and unconnected railroads and bridges, Vanderbilt had incorporated his new railroad, to be called the New York Central, all the way to Chicago. He also built a terminal, Grand Central Depot (1871), on Manhattan Island. All the other railroads (except the New York, New Haven & Hartford, which shared New York Central's facilities) had to ferry their passengers from New Jersey across to Manhattan. Therefore, freight had to travel from and to any one of the numerous New Jersey freight terminals over the Hudson River.

Before that depot, shipping ruled the commercial world of New York. The city as a collection of bridges, eventually ten, that crossed the East River awaited the first in 1883 and then not another until 1903. Across the Hudson, only the George Washington Bridge in 1931 and the major tunnels would not be finished until in 1927. Manhattan, Brooklyn, Queens, and Long Island, along with Staten Island remained landlocked well into the 20th century.

With no rail link to Brooklyn, Queens or Long Island, for some years freight had to be unloaded from the freight cars in Manhattan, and moved to lighters, ships made to carry cargo the short distances to the other boroughs. Since the majority of rail lines terminated in New Jersey, by the 1870s, groups of freight cars were able to be moved in groups using carfloats. Starting with a sturdy barge, railroad tracks were laid down on the craft side by side so that ten freight cars, five per side, could be pushed onto the barge in New Jersey. Then, with tugs pushing the barge, the carfloat barge traveled across the Hudson to an unloading and loading site in Manhattan or Brooklyn. (To this day, as viewed on YouTube, one can see one of the carfloats in operation.)

Once the cars docked at a specially designed "bridge," they are pulled off by locomotive and either unloaded or moved to be taken elsewhere. The cost and trouble this operation could cause was just one of the reasons for the Pennsylvania Railroad to finally have its own rail terminal in Manhattan at the end of the century.

By 1910, Pennsylvania Station, between Seventh and Eighth avenues and 31st and 33rd streets, featured two tunnels under the Hudson River, and four tunnels under the East River to Brooklyn, Queens, and Long Island to accommodate its newly purchased Long Island Railroad.

There remained problems with railroads. The old problem of gauges that did not match between cities still existed and it wasn't until 1883 that all the railroads could agree on what time it was. Some towns knew what

**Railroad cars are shown here being moved by water (April 1956). This practice, a precursor to container shipping, continues today. This use of a barge sometimes specifically built with side by side sets of rails capable of being moved by tugs even with the weight of a locomotive and freight cars seems to have begun in 1866. YouTube now shows videos of these carfloats.**

time it was by looking at any public clock, as in a church steeple or a display in a timepiece store window.

## *The Delays Go On*

The struggle at Flood Rock continued and intensified without sufficient funding. "The heavy ice running in the river rendered it impracticable to continue regular work. The pumps and machinery were, therefore, removed from the mine, and it was allowed to fill with water." But the water in the tunnels, all East River water, has a high salinity due to the rushing tides from the Sound and the ocean. To be cautious, non-salted water would have to found. "Considerable difficulty was experienced in obtaining fresh water for the boilers … owing to the freezing of the water in the lead pipe leading to the rock from the shore. One day's work was lost owing to an extraordinary high tide occurring, with high wind, which caused a large quantity of water to flow into the mine through the shaft; an entire day was required to pump this out far enough to continue work."[6]

Even though awaiting the appropriation to arrive, the delay did not mean General Newton was not busy in 1878. His supervisory duties included, for example, the "improvement" work on the Raritan River, Echo Harbor, and on the entrance into Charleston Harbor.

What work could be accomplished on Flood Rock in the tunnels already begun was confined to maintenance. The constant leaking of water in the work space always impeded progress, as it had at Hallet's and in the depths of Flood Rock Island's galleries and tunnels. Water flowed into the works from high tides, storms, cracks in the rocks, as well as from seams developing in the tunnels. It should not be forgotten that these tunnels had, as their roof, some little space of rock and then 40 feet of the East River above. The unused pumps would not be put back into operation to raise the water out of the works until July when workers could again be hired.

## The Harlem River Studied Again

At least part of the work concerning the East River received some attention on February 18, 1878, when Congress began to seriously consider "the Improvement of Harlem River, New York, so as to connect America's inland seas, Erie Canal, and Hudson River, by a short cut through the Harlem, with the Long Island Sound and the Atlantic."[7] After two attempts at funding had failed, a third lobbying effort came from the Committee on Internal Trade and Improvement of New York's Chamber of Commerce. Later an estimate claimed that the new route would reduce towing costs by 50 percent.

But survey work on the Harlem River and the impossibly narrow Spuyten Duyvil Creek on the Hudson (at about 227th Street) was funded to as much as $300,000. These two waterways have demanded attention for many decades for the sole purpose of shortening the passage to New England and to the Atlantic Ocean. Part of the desire for the Harlem River (flowing from 220th Street to about 100th Street in Manhattan) came from the fact that by 1878 more than 4,000 sailing vessels were arriving at the Port of New York each year, weighing a total of 1,087,439 tons. And the vessels grew in cargo size every year, even though sailing vessels still outnumbered steam by more than three to one.

For goods shipped by water from as far west as Duluth, the shortest way to the port of New York remained via the Great Lakes, the Erie Canal and so down the Hudson River. But if the goods from the West needed to travel further than New York but out into the Atlantic, building a shorter way from the Hudson to the ocean would save time and money. This would mean ships south from Albany could exit the Hudson into Spuyten Duyvil and then into the Harlem River down into Hell Gate. From there the ships could sail northeast into Long Island Sound, and thus into the Atlantic. Otherwise using the Hudson, an extra 20 miles was added to the journey to move down to the Battery, sail through the traffic in the port of New York, and up once more to Hell Gate. In 1877, Newton proposed widening the Bronx Kill passage between Randall's Island and the Bronx that would allow ships from the Sound to dock on the Harlem River.

# TWELVE

# Newton's Work Slows

## *Newton's Strategy*

Newton continued to point out the urgency of finishing the work as captains took their vessels through the East River.

The east passageway between Astoria and Flood Rock measured about 350 feet. There was some 41,000 feet between Blackwell's (Roosevelt) and Ward's Islands—with Flood Rock laying less than halfway to Ward's Island.

The space between bodies of land—from Queens across to 59th Street in Manhattan—measured 775 feet. But the width of the channel has nothing to do with the space between the two bodies of land. The depth varied within the channel, but once out of the channel (and closer to a shore) the depth of the water diminished by as much as two-thirds.

Clogging up an East River passageway as it moved north of Astoria were a series of rocks and reefs. In 1866, these seven obstacles formed a mass that had to be avoided—this led to the creation of the "Main Ship Channel" which was an attempt to bypass the seven masses of gneiss rock. Four strategies were implemented.

First, by using Newton's Scow, to finish the work on any remaining rocks not yet reduced to the 26 feet level.

Second, fill in the space between Great Mill Rock and Little Mill Rock with debris from various jobs.

Third, use a kind of "spring fender" where possible on some rocks simply to smooth the jaggedness.

Fourth, blow up Flood Rock, a nine-acre behemoth.

Most measurements said that the Hell Gate passage was at least 80 miles shorter than via Sandy Hook. Besides the waters of the Sound being much smoother, and therefore safer, less fuel was burned by the steamers. And while it remained true that fog at Hell Gate was a problem, the fog was even denser at Sandy Hook bar. At that time, no one even knew if shoals might permanently block the entrance to the harbor.

Even with the limitations at Sandy Hook, citizens and wealth continued to flow in and out of the port and "New York City is now claimed to have in its boundaries one-thirtieth of the population of the country, a city whose commerce alone pays one-third of the national revenue," wrote the *Times* on March 30, 1866.

First some clarification. Rocks seen above the waterline proved to be part of a larger reef, and only after exhaustive depth surveying was the extent discovered below the water. Flood Rock early in the 19th century is often referred to as a part of the Middle Reef. The other stony obstacles in the Middle Reef consisted of Hen and Chickens Reef, Negro Head, and possibly including Little Negro Head, another rock to be found on a map off 91st Street. No one piloted a vessel into the midst of the Middle Reef. No one seems to have measured its totality. Flood Rock was an area of about nine acres in the middle of the channels at Hell Gate. It showed in the river as having a small backbone projecting above the water, and it caught vessels swept upon it by the ebb currents, which passed directly over the rock. Newton knew the nine acres were 99 percent underwater, but only once the tunneling progressed would he knew how far the tunneling would have to go. The entire rock was going to have to be detonated and the tunneling had to cover the needed distance to encompass all of Flood Rock.

General Newton now fully realized that Flood Rock's destruction, the largest part of that reef at 392,040 square feet, would create a very large sailing area in the very midst of Hell Gate. In the old maps, Flood Rock, not joined but close to the rest of the Middle Reef, is shaped wider than it is deep, meaning it is likely that that wide portion takes up about 1,200 feet in Hell Gate.

This leaves a total of 400 feet on either side of it. Four hundred feet of sailing area becomes even smaller when one factors in the currents, swirls, eddies, tides, and whirlpools awaiting a vessel, all ready to shove a ship onto the Middle Reef or against Little Mill Rock.

Blowing up Flood Rock would, in effect, form a brand-new channel through the East River, to begin with, as wide as 400 feet. With activity already in progress, the other parts of the Middle Reef—The Gridiron, Hen and Chickens, the two Negro Heads—would be broken and dredged. Where there was a Middle Reef nothing would remain 30 feet deep into the bed of the East River. View an NOAA map in 2020 and look in the very middle of Hell Gate, and that 30 foot marking is apparent.

## Progress at Flood Rock

Much remained to be done. Unlike the Hallet's effort, workers on this project could only reach the site by boat, though the boat departed from

Hallet's Point. Workers at Flood Rock had a very clear view of what river pilots called The Turntable, another whirlpool that held boats as if they had dropped anchor. And they had to contend with more problems at Flood Rock. On January 14, 1878, the steam boiler brought from Hallet's Point was judged to be too weak to power the drills. Drilling was not resumed until February 14, 1878. By being assured of a July 1, 1878 appropriation of $350,000, on June 26 four Burleigh drills were able to be put back to work where they would do the most good.

The pumps on Flood Rock were started July 8, 1878, and by July 14 had removed the water from the mine's tunnels, galleries, and piers. It was then found that during the 18 months the mine had been full of water, numerous seams had opened in the roof and sidewalls, which required timbering to prevent the piers and roof-rock from collapse. No work except timbering was undertaken in the tunnels until July 31, 1878, when one drill was started, and on August 1, two more. These were kept at work until September 17, when a fall of rock occurred, seriously injuring two men and rendering the planked gangways so unsafe as to prevent any further work. Laborers were instructed to secure the walls by timbers.

Two pieces of information that described events in Newton's time can prove illustrative of the kind of operation Newton commanded, supervised, and contributed to. One contrast has been the safety record with handling explosives. The only accident occurred on September 4, 1876. Readers may remember how every year there were terrible and massive explosions elsewhere, but not at Hallet's nor at Flood Rock. In January 1878, there was another accident at the Hoosac Tunnel. A watchman named Wilson died and the factory was demolished in another blast, this time caused by the so-called "1879 compound," the George Mowbray improvement that has always been regarded as being perfectly safe. Worse, the explosives were in a cold or frozen state, it being Massachusetts in January, during temperatures when dynamite and nitroglycerine were least likely to explode.

In 1878, Newton could report that "the work at Flood Rock was prosecuted more energetically, but with varying appropriations." His superiors were used to reading his reports which often included his accounting even down to each foot of wire used in his work. This was the Gilded Age in which the wealthy had few restraints on their spending. Newton as an Army officer was spending federal government money. In the city in which he lived and worked, a famous city government building project had been going on since 1861: the Tweed Courthouse next to City Hall. Tammany Hall, applying their use of kickbacks, payoffs, cost inflation, and corruption, would see to it that the five-story building covering 177,500 square feet would end up costing $11 to $12 million. In contrast, the LeVeque Towers in Columbus, Ohio, 42 stories taller and 200,000 square feet bigger, cost just $8 million and took three years to finish. The Tweed Courthouse took 20 years.

Many decisions remained for General Newton and his staff. After calculations for cost had been made, it was decided to limit the amount of "mining"—drilling to make wide galleries and tunnels—that was both time and money consuming. "Galleries were considered only necessary to give access," Newton's report read, "so as thoroughly to shatter the rock."[1] What also remained was the expansion of the capability of the Flood Rock work area.

This East River work continued to be as important to commerce as the size of ships began to increase. For example, in 1878, a sailing ship could no longer compete effectively with the steamship for the China trade. Iron ships proved to be "30% to 40% lighter and had 15% more cargo capacity compared with wood armature ship of the same size."[2] An iron ship could accurately predict the time for the delivery of goods. Setting a new record in 1878, a ship's crew could brag about spending just 90 hours to move freight 1,300 miles from Havana to New York City. The 14 miles per hour speed might be slower than a trim clipper, but the iron hold was bigger and with the implementation of screw propellers the ship did not rely on the vagaries of wind and would be better able to handle storms.

As the ships grew larger, newspapers echoed the cry heard for decades, that the city was using about 4,700 feet of the East River's Manhattan shore of dockage and that only reached Chambers Street. To have ship find berths even just to 14th Street would allow another 9,000 feet of space for piers. Even the Hudson shoreline in Manhattan did not extend above 23rd Street at the time. And now with the East River opening up at last, even more vessels will need dockage and many writers complained that the commerce of the city will be stunted. Even 12 years before October 8, 1866, *The Herald* worried "When prosperity returns, industry and commerce revive, and our harbor fills with ships from all quarters of the globe, we shall be unprepared to accommodate them, and shall inevitably lose a part of that on which we depend for our increase in the future."

Early in 1879, March 3, the Hell Gate project was just months away from the end of its $250,000 appropriation from fiscal year 1878–79. It remains uncertain if some of this money would be used for projects on the Hudson, but what is certain was how vigorously Newton used this Hell Gate funding while he had it to hand. Much of the end of 1878 and all of 1879 would be used to establish the installing of equipment, mining the tunnels, and more efficiently disposing of the rock rubble.

Some of the appropriations went to try to finish the work on the river with the Newton scow, sites like Shelldrake, Frying Pan Rock, and Way's Reef. The problem of the lowering of Diamond Reef went on as it had for many years.

The dredging at Hallet's Point too continued. But a considerable amount of money was going to be spent at Flood Rock since work which had begun in July 1875 had been mostly suspended since January 1876. Flood Rock, that huge mass of gneiss 1,200 feet from the Long Island shore, covered an area of about nine acres on the bottom. But only a small portion, 70 feet by 250 feet, was all that showed above water. (A brick magazine for storing explosives had briefly been installed on Flood Rock in 1870.)

Meanwhile, the British, thinking of their ships, made it clear that the East River, according to the British journal *Marine Engineer and Naval Architect* "was the most direct road between New York and Europe."[3] For Americans and New Yorkers, in fiscal year 1878–1879, the "direct road" contributed over $98 million in customs duties to New York's collection district. Adding up the value imports and exports at the port of New York totaled $564,000,000 in 1879.

By 1879 two vertical shafts 65 feet apart were completed on the expanded surface. The main shaft was completed to hoist up blasted and drilled rock; the smaller shaft was used briefly for the tubes bringing the compressed air to the drills. As the shafts were being finished, Newton began to think once more of how to use the most economical means for the massive work he was undertaking.

Promoted to full colonel status in the regular army on June 30, 1879 (while maintaining the title of brevet general of volunteers), Newton had to find a workable and efficient method of how to take the debris out of the shafts and tunnels, as Captain Mercur wrote in July 1879, and dump it nearby, principally between the two Mill Rocks, about 400 feet away. The General remembered the cost of hauling large pieces of Hallet's Point, an amount of rock equaling 234,900,000 pounds. In 1875 Newton had calculated the average cost of dredging and dumping one cubic yard of debris, to be $4.29. So, to remove the Hallet's rock cost $373,230.

## Expanding the Site

One of Newton's problems was that at the beginning of the project only 250 square feet of that rock island reef could be seen above water. Therefore, he ordered that for about five months in the beginning of the year more debris be "used for building an island large enough … for carrying on the work and so placed as to serve as a dike for deflecting the ebb-tide."[4] This deflection made it easier to dock the scows used to carry off debris. Flood Rock Island would also have the effect of diminishing and moving a whirlpool off nearer to Astoria. The artificial island rose to about seven feet above high-water.

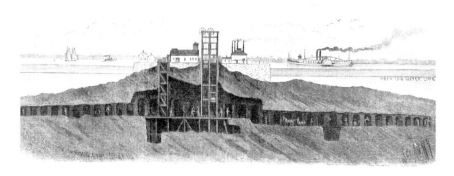

**Rear and cross section view of the mining of Flood Rock. The artist chose to show tunneling and shafts, the stairs on the left, the two shafts for raising the debris and returning empty. A mule can be viewed pulling drilled rock at right center (*Scientific American*, January 2, 1886).**

As for machinery, the new artificial island could now allow for "a solid and ample place for the machinery, sheds, boilers, pumps, derrick, etc." Put into place were "boilers, a pair of hoisting engines, one large mining pump, one circulating pump, one large surface condenser, and a number of drills." Added to the new equipment were "a head frame and cages for hoisting" while "a boiler-house, coal-bin, and engine-house have been built."

*Manufacturer and Builder* magazine of September 1879 thought "in addition to the work done on the mine, permanent arrangements have been made for carrying on the work." One of these schemes included the building of a small powder-shed, for the storage of the cartridges of dynamite for immediate use, while the majority of the explosives were isolated, several hundred yards north on Little Mill Rock. Listed among the 70 men employed in the day shift were engineers, drillers, blacksmiths, and laborers. "Typical of Newton's thrift, the machines used at Hallet's Point to service, restore and support the major tools—the pumps, the compressors, the drills—have been thoroughly repaired to serve again at Flood Rock Island."[5] The General's plan was to distribute the explosives in such a way that when the blast is over little or no dredging will be necessary but rather the debris will rest in the crater made by the blast in the river's bed.

So before very long, Newton's "plant" could look like the drawing above from Manhattan. One notable addition to the work is the mounting of stairs in shaft number two.

But how to lower costs had always been one of Newton's concerns. And always part of the cost would be speed. What had he learned and used at Hallet's? What alterations and improvements to those methods might be made? At Hallet's, Newton was able to use a locomotive above the pit to move debris, the small train's cars filled with debris raised by the derricks.

The locomotive pulling power could be used to transport the mined material to a dump some small distance away from the works in another part of Astoria perhaps near Pot Cove. But no such pulling engine of a steam locomotive could be transported and lowered. Besides, at Flood Rock, the drilled material must be taken off of the nine acre obstacle and deposited where it might serve some purpose. Likewise, Newton saw to it that Flood Rock Island was so equipped to unload coal used to fire the machinery.

The first part of Newton's solution required mules and tracks. Here the pulling was a matter of harnesses, not locomotives, so a savings there. He first laid down 2½-foot-gauge tracks. Next lowered into the shaft were the cars and their tracks with mules to pull the loaded cars. A space was created to stable the animals, now living in the shaft for more than two years. The tending to the mules included feeding, watering and disposing of their waste. These tipping cars worked in much the same way as a modern day dump truck. Once filled and level, they can be tilted so that the contents slid out.

To the job of disposing of the rock rubble, Newton would expend funds for the construction of wharves on Flood Rock to accommodate scows, "the necessary chutes and pockets for passing the stone from the cars to the scows, and the completion of the self-dumping cages and necessary mine cars, together with the purchase of a powerful tug and two dumping scows for removing the broken rock."[6]

The idea of "self-dumping cages" was further explained later this way: "The descending empty car counterbalances an equivalent weight of the ascending load in the adjoining lift, thus practically eliminating the cost of hoisting the cars as dead weight. The loaded scows are taken by a powerful tugboat to the dumping ground."[7]

In this way, the drilling is done, the cars are loaded, the mules pull the cars to the center of the shaft and the cars are pushed into a hoisting cage. Then, the cage is brought up to the top of Flood Rock, the car designed in such a way that it tilts. These cars were variously called "tipping Cars" and "self-dumping cages." Awaiting the car's load was an iron chute aimed at a dumping scow alongside the rock.

As was so often true with Newton in command, what had to be considered was the time (and money) to raise the rocks from Flood Rock, the time to move the rocks, and the men and machinery he had to contract with.

Thus, with Flood Rock Island more fully shored up by the rubble, with some wharves added and with the purchase of a powerful tug along with the two dumping cars, the debris could be moved more rapidly. The dumping cars can be hauled up out of the shaft, and tilted onto the new chute, which incline lets the rock move directly onto a scow at the new wharf. We could see the wall around Flood Rock, the waiting scow, the five buildings on Flood Rock Island, and more. And when Flood Rock was demolished, the tugs, the scows, the Newton steam drilling scow all would be put to work on the Harlem River and other places as well.

While it was true that the General had planned, over the years, to purchase a small fleet of tugs, scows, and smaller ships and in that way avoid having to pay others for services, each scow, for example, might be able to handle less than a cubic yard of rocks. Even so, given that the appropriation was forthcoming, "With these facilities it is proposed to increase the present rate of working about four times, or more," Newton reported.[8] Of course, not everything went smoothly. For example, in January 1877, "workmen were said to have been imprisoned" for a day by the ice and unable to leave Flood Rock.

## Newton Considers

Elsewhere, instances of the danger of high explosives appeared often in newspapers. An explosion at Stratford, Ontario, on May 5, 1879, apparently on a freight train, killed two men and wounded several, while fifty cars were "reduced to match wood, and as many houses wrecked."[9] Once more at the Hoosac Tunnel, now a storage area and not a construction site for the railroad, "Two Men Blown to Pieces; North Adams's Fifth Explosion…. The packing house blew up and William Long and James Pierce were 'torn to atoms' by the explosion," the *Times* of July 1, 1879, reported. "Fragments of their bodies had been scattered over an area of several acres. The largest single piece of either man that was found would weigh about 30 pounds, and that was the trunk of Long's body."

Back in New York City, the Flood Rock tunnels that General Newton designed were carved out of solid rock. The engineer Colonel DeWitt Clinton Haskins had decided in 1873 to try to dig a tunnel across the Hudson River, beginning on the New Jersey side devising a way to deal the difficult river bed The tunnel, a first attempt to build a rail tunnel under the Hudson River, would not be finished during Haskins' lifetime. Blowouts, the end of funding for the project, and the difficult river bottom soil meant that the future PATH tunnel would not be finished for 35 years, until 1908.

Now being paid $3,500 at the rank of colonel, Newton was given more responsibility as a member of the Permanent Board of Engineers for Fortifications, River, and Harbor Improvements. It may be that his responsibility increased when his brilliant assistant, Julius H. Striedinger, left New York unexpectedly in the year 1879 or 1880, and went to South America under a three year contract with the Government of Colombia, "to make a new exploratory investigation of [unnamed] river and its problems."[10] Even though we do not know when he returned, we know he was at the 1885 explosion.

(A contributor to the literature of blasting, of demolition of buildings, of detonation, of fuses, of underwater lanterns, and even of the control of fires, Striedinger died sometime before 1899. His contribution to the reconfiguration of the East River can be measured, in part, by the 43,000 tons of rock dredged by the Hallet's Pont demolition from 1877 to 1879.)

At Flood Rock 1,307 linear feet galleries were driven, thereby removing 5,115 cubic yards of rock. At five rocks in the river, the Newton steam drilling scow continued its work of making those obstacles safer for navigation.

With the new bridge towers now in place, beginning to draw even more attention, the East River (now Brooklyn) Bridge was so much bigger than the Flood Rock project in so many ways. Maybe not larger in importance, but certainly in scope and visibility, the Flood Rock work being 95 percent underground. The towers themselves, higher than any building at 276 feet, were massive and the roadbed raised to be 135 feet above the East River. More than 600 workers worked at completing the bridge, swarming over it all day, while about 70 labored under Newton's command beneath the surface.

Perhaps as a measure of the growth of the wealth coming into New York Harbor, 1879's vessels can be tabulated. Of arrivals from foreign countries, 7,348 were counted. Using Hell Gate alone as the way into New York Harbor, from United States points north and Canada, 8,323 ships were tallied. From these 15,671 vessels in 1879 the value of imports and exports reached $564,000,000. It seems that as traffic increased, as commerce expanded, as the wealth of New York City became more apparent, so did the reportage on Newton's nine-acre Flood Rock Island appear both in frequency and in detail. If more and better dockage were to be built, Newton was told, then fewer ships will have to wait to get into port. The White Star Line, for example, moved from New Jersey to New York City.

Newton's work on the nine acre Flood Rock would take less than a decade. As one other contrasting engineering project, there was the problem of Elderslie Rock beneath the River Clyde near Glasgow, Scotland. Not discovered until 1854 when the seven-acre boulder severely damaged the steamship *Glasgow*, the rock was worked on for 53 years before it could be lowered to a 28-foot depth so as to accommodate the passage of large ships.

*The New American Cyclopaedia: A Popular Dictionary of General Knowledge* said, "There is no actual record of the coasting trade of New York, but it is certain that there were not fewer than 9,500 arrivals coastwise and an equal number of clearances."[11] What is known is the tonnage of "coastwise and Internal Trade" held steady at between two and a half million to four million until the end of the century. The "coasters" entered New

York Harbor through Hell Gate. All of these vessels might be combined with unrigged sailed ships, such as scows, barges, and canal boats to give a more complete portrait of customs duties as New York was approaching $130 million in 1880. Excise tax collected for goods from inside the U.S. reached $124 million. Also in 1880, New York City, now with 1.2 million people, saw steam tonnage rise to over four million. By then steel had mostly replaced iron in ship building.

Life in the city in 1880 included more than 150,000 horses, a number which would rise in the next few decades. Used for all manner of pulling tasks, these harnessed animals delivered passengers to ferries and docks and took them and their luggage away. These work horses produced 20 pounds of manure and 2 pints of urine per day. The manure flooded the market to such a level that farmers were paid to take it. Piles of manure were often 50 feet high. Dead and rotting horses littered the streets. All this attracted massive numbers of flies which spread typhoid fever and other diseases. Horse-drawn vehicles killed people at far higher rates than today's vehicles.

Even as Newton's projects went forward, steam trains had reached speeds of about 38 MPH, as railroad mileage nationwide hit 163,600 miles with nine Eastern states accounting for 17 percent of the traffic.

## Interest Increases

Throughout 1880 reports in newspapers and scientific journals began to print more specific details of the work on Flood Rock, work which went on 24 hours each working day, employing 135 men in three shifts. A report from *Scientific American* of October 16, 1880, reads, "the whole excavation will resemble an immense cave, the roof being supported by the rocky pillars which now form the sides of the headings [tunnels]."[12] The cave-like digging by then had produced 20 horizontal passages running in one direction and 11 headings which have been dug across to meet these 20 at right angles. These tunnels could be as high as seven feet and as wide as ten feet.

About a third of the tunnels could in 1880 be said to be complete. Extending out from the main shaft, the three acres so far completed had ceilings (or roofs) about 25 feet thick but eventually only ten feet would separate the workers from the bottom of the 35 to 40 feet deep East River. Newton would later explain in the February 1886 *The Popular Science Monthly*:

Considerable risk was incurred in this part of the work, from the danger of the rock crumbling, and from the uneven and uncertain thickness of the roof. The average thickness was 18.8 [feet] thick, and a minimum thickness of ten feet. The exact thickness could not be ascertained beforehand, for no soundings could distinguish between the solid rock and a concretion of bowlders and shells formed upon it."[13]

Until 1880, the excavation of tunnels had been slow work—the material being mined is gneiss—but Newton's report could read that "24,000 cubic yards of rock were removed, 43,000 blasts made, and 57,066 drills sharpened. The number of blasts made each night now averages 150."[14] The General could take full credit for the work that went on in the river—the removal still of 15,195 tons of broken rock from the Hallet's Point explosion, rock and "bowlders" removed from Frying Pan, Shelldrake, and Diamond reefs—as well as almost 13,000 cubic yards of rock from the "galleries" in Flood Rock.

His achievements may have led to his being appointed "Superintending Engineer of River, Harbor, and Channel Improvements, Surveys, etc." in and near New York Harbor in the period June 17, 1880, to April 1, 1884. Articles in newspapers joined the work of magazine writers to report on the progress. Newspapers reported miners, mechanics and laborers at work, using "from 20 to 80 drilling engines … making 'holes something over two inches diameter' … each about four feet deep, at the rate of 81–93 feet per shift, by each active drill."[15] Still alert to the public's unease of his work with powerful explosives, Newton decided on having the ordinary blasting done at night. In its September 24, 1881, issue, *Scientific American* reported hard facts: "as many as 800 holes have been fired … in a single night." To keep the tunnels and shaft clear "the ventilating fan, located at the top of the shaft, is run at its maximum rate, displacing about 50,000 cubic feet of air per minute."[16]

The Flood Rock project by 1880 had become so well-known and Newton's reputation so well established that the visiting Count de Lesseps asked to visit the site. De Lesseps, who had so much to do with the construction of the Suez Canal, and who was then in New York City to raise money for the building of the Panama Canal, was taken by tug to Flood Rock for an inspection tour.

## Other Projects at the Same Time

After Newton's many successes making navigation on the East River safer, the nation's continuing and growing reliance on the port of New York underscored the need for many projects that would continue to increase the effectiveness of New York harbor.

The total length of galleries on June 30, 1881, was 13,523 linear feet and 21,528 cubic yards removed.[17] Throughout the year, the dredging of Hallet's Point and the removal of debris from the Flood Rock works—about 81 million pounds of rock—in addition to more than five million pounds from the rock called Heel Tap and the reef at North Brother Island—all would bring about smoother sailing on the East River. Along the river at Buttermilk Channel, widening work went on to facilitate the movement of vessels into Brooklyn, along with deepening projects at Gowanus Bay, Red Hook, Gowanus Greek, and Bay Ridge channels.

On March 3, 1881, a long-awaited development seemed to move forward. When Congress passed the River and Harbor Act (with the aim of building the Harlem River Shipping Canal) yet more surveys were ordered. Newton had made many measurements by this year, aiming at the construction of dockage on the river and building a link between the Hudson and Harlem rivers, providing a speedier route for ships into the Long Island Sound. *The New York Herald* of October 2 saw the construction as a way to "make transportation in New York Harbor near the confluence of the North and East rivers safer and more rapid by giving to the canal-boats, freight flats, tugs and the small fry of the river trade a better passage about New York City than that around the Battery." And it provided a faster route for ships into the Long Island Sound. This passage would not, however, be begun until January 9, 1888, nor open until June 18, 1895. (Eventually, the river's width would average about 400 feet while Spuyten Duyvil Swing Bridge spanned more than 600 feet.)

Around the same time, with prophetic vision, John Wolfe Ambrose recognized in advance of his fellows the danger of New York being handicapped through the inability to supply port accommodations to very large ships which within a few years would surely be built. When in 1881 he first went to Washington to ask appropriations for New York Harbor, Ambrose could cite statistics showing the high, and rising, percentage of vessels that needed to dock in the city. In a very short time, Ambrose would push for a channel from the Atlantic into the Narrows. For now, the East River remained the best choice.

Traffic continued to grow. It is known that 3,867 ships, steam and sail, used the harbor and uncounted towed barges still were coming down the Hudson. There was shipping too from Flood Rock's wharves. Scows, laden with debris, carried the rock 800 feet to the space between Little and Great Mill Rocks, to build the western side or breakwater to the new channel formed by the eventual removal of Flood Rock. This scheme was first proposed by Newton in 1866, and it was calculated that the removal would lessen the current around joined Mills Rocks from 9 knots to 6.

## The Working Force

The approach of the March 16, 1881 *New York Herald,* unlike the statistical writing of *Scientific American,* focused on the labor force who had already drilled and blasted five acres beneath the river. The paper took note that "the steam power must supply the energy for ... hoisting the debris" and mentioned the large boiler and feed pumps without neglecting the tasks "at the bottom of the shaft—drilling, ventilation, draining the drilling floor, where shifts of about sixty men each worked in less than the best conditions. ... In a rubber hat, a coat and boots to protect them from the constant dripping. the men try their best to ignore the sounds of the rushing cataract above their heads, as they pick their way through the gloom with miners' lights and hand help lamps." Above the tunnels, the galleries, blacksmiths repaired equipment, built tools, sharpened drill bits, all the while maintaining the boilers.

At this time *Harper's* told its readers more about the drying chamber. "Here the men of the several shifts changed their clothing ... leaving their water-soaked garments hung on long frames to dry. Nearby this mine house was the stable in which the three mine mules, Dan, John, and Mary, spent their leisure hours."[18]

## Slowdown of 1882–1883

Newton found himself in the midst of the 38-month Depression, the one lasting from March 1882 to May 1885, the third severe depression in the late 19th century.

Six years after the work on Flood Rock began, the Congress granted an appropriation to New York for $200,000 to continue the excavations at Hell Gate. But yet it is known that Newton suspended all work from July 20 to August 20, 1882, certainly for the work specifically on Flood Rock. It is known that on June 19, 1882, work was resumed at least on the river, work on the troublesome Pot Rock north of Pot Cove in Astoria. But even those tasks were halted when on August 27, Newton's steam drilling scow was crashed into three times, twice in one day. Seeing the damage to the scow and alert to the loss of the chains and anchors that held the scow in place, the tugs *Star* and *General A.A. Humphreys* towed the scow to a place of safety in Hallet's Cove. These collisions, while stopping its work, did no great damage to the vessel.

Once towed back out on the river, Newton's scow suffered ten more collisions during the short time back on Pot Rock, from June 19, 1882, to September 4, 1882. Moved to Frying Pan Reef on September 5, 1882, the scow then drilled, blasted, and removed broken rock until January 13, 1883, when it became necessary on account of the floating ice to go into winter quarters.

The national economy might have been weak but the customs revenue for fiscal year 1882–1883 for New York reached almost $148 million. As a contrast, the federal budget for 1883 amounted to $309 million, but with a surplus of $135 million. The halt in Flood Rock work in 1882 was repeated throughout 1883, awaiting the new fiscal year money on July 1, 1884. A report by one of Newton's staff, Lieutenant George McC. Derby, clearly stated, "all work was suspended by your order from July 20 to August 21, 1882, and from May 31 to June 30, 1883; and from March 31 to May 31, 1883, the work was carried on at one-third the capacity of the plant."[19]

Perhaps with the work in abeyance, Newton had time to do some scholarly work. *The North American Review* printed his nine page piece titled "Modern Explosives" in its November 1883 issue.

With the River and Harbor Bill failing to be approved in March 1883, Lieutenant Derby concluded "the progress reported is, therefore, rather less than nine months' work. These interruptions and delays are greatly to the disadvantage of the work, as the expense of pumping and superintendence goes on whether any progress is being made in the mine or not."[20] Money was unspent on coal for the boilers, on contracts for hauling debris, on workmen, with only a few watchmen still on site on Flood Rock Island.

Derby's report does show some work being completed but with severe limits. One positive item included the fact that "the leakage at present only amounts to 737 gallons per minute at half tide." But three large tasks remained. The thinning of the roof of the tunnels (called "stoping down"), the drilling of the large holes for the final blast, as well as "the locating and plotting of the [smaller] holes drilled for the final blast."

No doubt a slowdown did not attract much scrutiny from newspapers and magazines. Certainly in 1883 the excitement about the opening of the East River Bridge—the Brooklyn Bridge—captured a good deal of the public's attention. Work on the bridge, designed in 1867, lasted 16 years and at 135 feet high it allowed pedestrians a view of both New York and Brooklyn that few had ever seen. It had the advantage of being able to accommodate carriage, rail and foot traffic.

The bridge may have cut back on the business of some of the ferry companies—in 1883 there were 12 ferry routes using 10 different ferry terminals in Brooklyn and 11 in Manhattan—but it seemed that travel by water still dominated the life of New York City.

The country could claim that there now ran 600 railroads over 93,295 miles of track, but except for the New York Central Railroad all other Manhattan destinations and departures required taking ferries to begin or complete a trip. Even Long Island Railroad passengers would have to ferry across the East River. Travel by land was still slow and crude. The invention of macadam was really the installation of crushed stone. Employing Portland cement for roads remained a decade away. In the city, if more than 40 horses per day were dying on city streets, and the city still functioned, we can only estimate the number of horses and the cars they pulled.

But more than 10,000 ships entered and left ("cleared") the Port of New York each year and maps of the time show how crowded the harbor could be with rowboats, oystermen, fishermen, ferries, along with lighters that had offloaded cargo from ships at Sandy Hook.

General Newton, now engineer in charge of the improvements of rivers and harbors in the New York District, expressed an urgency, an impatience, with his Flood Rock project while supplying well thought out arguments. He insisted that fall was the only suitable time to explode Flood Rock. He explained that "the winter temperatures caused cold affecting the state of the dynamite," and likewise "in spring and summer," as Newton explained, "there is danger of electric storms, which, after the wires have been introduced ... will render an accidental explosion possible."[21] All these delays in completing the project meant additional expense because "the expense of pumping and superintendence goes on whether any progress is being made in the mine or not."

The general could point out that the raising of the debris from Hallet's Point was complete. Although he may not have known, the time for the building of steam-powered ships with auxiliary sails was over and, though the economy continued to suffer, in New York harbor nearly 70 percent of U.S. imports were moved.

At year's end came a notable death: General Humphreys died December 27, 1883, at age 73. Retired since June 30, 1879, Humphreys was succeeded by Horatio Wright who himself retired as chief of engineers on March 6, 1884.

## Newton in Charge

By early March 1884, Newton was notified he was about to be named to the post of Chief of the Corps of Engineers and promoted to the rank of brigadier general, with a pay of $5,500. After April 8, 1884, Newton would now be living in D.C. At this point, the reporters learned something of his personal life. He would suspend his membership in the Manhattan and Century clubs in New York City, and also, being a Catholic, in the Xavier Union. This last group "is purely social in its objects, and ranks among the foremost of the fashionable clubs...."[22] Later renamed "Catholic Club of the City of New York," the group gathered at 267 Fifth Avenue.

Newton's assistant on site was Captain James Mercur who supervised the work of Lieutenant Derby. But it seems true that no work on any consequence, other than maintenance of equipment and the tunnels, took place between May 31, 1883, and July 1884, when $360,000 was allotted to the Hell Gate project. Under the Lieutenant, some savings were accomplished by having the Flood Rock pumps work in tandem to save energy and the price of coal. Then too the leaks at Flood Rock had been reduced to 565 gallons per minute. But the 200 or so drillers and miners had been absent for 13 months except for specialists to monitor the continued threat of flooding. "One seam encountered was 10 inches wide and 100 feet long; and another, from 1 to 4 inches wide, extended right across the reef, over 400 feet in length. They were filled up with Portland cement as they were opened out." Harcourt noted. "Considerable risk was incurred in this part of the work, from the danger of the rock crumbling, and from the uneven and uncertain thickness of the roof."[23]

Other changes went on as ancillaries to his taking command. Newton was selected as a member of the Washington Monument Commission (to be opened in 1886), as a member of the Light-House Board, and of the Board for Determining Positions for Atlantic and Gulf Interior Coast-Line Defenses.

Even though it was written that "During the year ending June 30, 1884, the available balance will be sufficient only for the care and preservation of the work and running the pumps to keep the excavation on Flood Rock from filling with water," Hell Gate and other places on the East River needed work.[24] The Hell Gate project was enlarged even further after the passenger steamer *Pilgrim* struck an undiscovered obstacle on May 4, just off Manhattan's 19th Street. Named Pilgrim Rock, its removal was added to projects already underway, such as at the troublesome Diamond Reef. Scheduled to be future projects after surveys were finished were Ferry Reef and Charlotte Rock, both opposite 34th street; the removal of Shell Reef off 9th Street, and a reef off Sunken Meadows (then separate from Randall's Island); and Baretto Reef (at Baretto Point in the Bronx, opposite Rikers Island).

To protect shipping on the East River, the year 1884 saw erected what was "The Most Powerful Light in the World." Powered by a 23 horsepower engine, nine lamps each produced 6,000 candle power—equal to 46 light bulbs of 100 watts. To see the light at Astoria shine for the first time, "crowds gathered ... to witness the effect." One part of the ring of lights brought "into bold relief ... the great public buildings on Randall's, Ward's, and Blackwell islands." (Two years earlier, Edison's Pearl Street Station, an electric power plant, supplied electric power and light to a small part of southeastern Manhattan.) The Astoria 250-foot-high light lasted just four years.[25]

In shipping, nearly 70 percent of U.S. imports were coming through New York, while 11,538 major ships were counted as they entered and exited the port. By 1884 the end had come for the building of merchant ships with sails, even for steamships with auxiliary sails. Even though "America led the world in maritime enterprises by the mid–1880's," Ms. Rattray writes, "and the Port of New York led America" leading even to taking in $135 million in customs duties, "America was still behind the rest of the world in the ownership of steam-powered vessels."[26]

Once the appropriation money had been deposited in New York, two major tasks remained for the rest of the calendar year. First, the galleries had to be completed, and second, those tunnels, roofs, and the piers holding up the roof had to be prepared for the final blast with the drilling of "about 4,800 feet of 3-inch holes to be drilled in the roof," Captain Mercur wrote, "and all of the holes in the pillars; about 6,500 cubic yards of broken stone to be removed; and about three-quarters of an acre to be tunneled and completed, provided the water does not prevent our reaching it."[27]

With the money finally becoming available in July 1884, the destruction of Flood Rock would still be unfinished in July of 1885. Newton's choice of having the explosion in the autumn would hold.

# THIRTEEN

# Newton Readies the Explosion

## Strategies Are Decided

Should any doubt remain of the vital commercial aspect of Newton's work, for decades until 1913, the Port of New York at its Custom House at 55 Wall Street "collected an astounding two-thirds of the federal government's revenue."[1] Newton was well aware of the commercial aspect of the destruction of Flood Rock. He writes, "the interests at stake were so great, and the details to be looked after to avoid every chance of miscarriage so numerous" that he would spend hours even on the day of the detonation seeing for himself that all matters were in order.[2]

The traffic in the harbor increased again. Fourteen million tons of vessels arrived and left the port of New York in fiscal year 1884–1885. Demonstrated here was a lack of balance between the receipts of the New York City Custom House—more than $126 million—and the absence of funds for the Hell Gate project as of July 25, 1885.

In 1885, with the use of Newton's steam drilling scow, Pilgrim Rock (found just the year before) was demolished down to 24 feet depth. Money had been spent on experiments with explosives, the most of which was to find the least expensive way to demolish Flood Rock. The thrift, the use of the taxpayers' money, that had always been insisted on by General Newton, became even more important when it became known that the appropriation of 1885–1886 was zero dollars, and no money would be appropriated for 25 months—from July 5, 1884, until August 5, 1886. But by mid–July 1885, Newton saw to it that the prepared cartridges of the nitroglycerine compound were to be set in subdivided wooden boxes, each of which held a dozen.

Still well-hidden, Flood Rock may have appeared as tiny on charts and in normal navigation it would be easy to miss since so little of it showed. In addition, the nine acres lay about midway in the space of 1,292 feet between Manhattan and Astoria.

On Flood Rock, the project was being supervised by Mercur and under his leadership, McC. Derby. Derby's focus would be on the explosives themselves. By year's beginning the whole nine acres—392,040 square feet—of the reef was cut through with tunnels, those "galleries" supported by piers reaching up to the very bottom of the East River.

It is unknown exactly how much General Newton had to do with the day-to-day operations as late as 1885, but it is known that the overall plan remained very much important to the General and that he had many years to be able to see what the final steps would have to be:

- Sketch out the placement of drill holes for complete and maximum effect
- Decide which holes will contain the primary charges—591—and which the secondary charges (21 percent)
- Drill the holes
- Transport the chemicals for the project
- Make the explosive
- Fill the cartridges with the explosive compound
- Transport the cartridges from Mill Rock's mixing house to Flood Rock
- Move the cartridges from the transport vessel to the shaft.
- Take the cartridges down to the tunnels and piers
- Insert the explosives
- Secure the cartridges in their holes
- Attach the fulminate of mercury
- Connect the wires to the explosives
- Run the wires through the tunnels, up into the river and take the wires to batteries on shore
- Set up the prime battery on shore

Newton's design counted on 20 percent of the cartridges being connected to a battery, with the other 80 percent of the cartridges to be detonated by "sympathy." Newton knew what he wanted to do. "Lift the whole bed of rock right up from its resting place."[3] The plan of making the excavations large enough to swallow all the debris of the reef and leave a channel deep enough, without further operations, was discarded by Newton. In coming years with the work finished and the blasting and dredging accomplished, Newton knew a very broad channel would have been made in the middle of Hell Gate. He insisted "If the funds are supplied as needed, the dredging can all be completed in three years" and the depth of 26 feet can then be achieved.[4]

By the end of the drilling of holes at Flood Rock—the making and placing of holes where explosives had to be inserted—Newton's report counted 11,789 drill-holes in the roof and 772 in the pillars, and the holes' total length altogether was 113,102 feet, or more than 20 miles. As for the tunnels, their total length amounted to 21,070 feet, at a depth of 50 feet below the low water level. The tunnels were at an average 10 feet from floor to ceiling, and 6 or 8 feet wide. Already the reef had yielded 80,232 cubic yards. That is, broken rock taken from the reef during excavation weighed more than 216 million pounds.

As the time for the final explosion drew closer, in this case August 12, 1885, Newton once more pointed out the business aspect of his work. "The amount of commerce and navigation benefited by the completion of this work," he wrote, rose to "about $4,000,000 daily."[5] This estimate becomes even more significant once the 1882–85 business downturn was calculated at minus 32.8 percent. Sources other than Newton agreed. According to the *Harper's* of October 16, 1885, "The value of vessels and their cargoes passing Hell Gate each day is estimated to be between $4,000,000 and $5,000,000, and as this vast traffic is constantly increasing, the importance of removing the obstructions at this point can readily be appreciated."[6] This statement was made even as the railroad miles in the country had passed 100,000.

Flood Rock Island had become a place dotted with a derrick, a water tower, wooden buildings; and it was noisy, with clattering from the crash of rocks sliding down a metal chute and booming nightly explosions. New York newspapers began to pay very close attention as Newton's projected date drew near. An article in the magazine *Harper's* was supported by 11,000 words and six illustrations, using a first person narrative for a reporter's experiences.

Once the writer disembarked from the tug *General A.A. Humphreys*, he "was provided with a helmet, cape, and huge boots of rubber, and a little flaring oil lamp attached to the end of a two-foot stick" to outfit him the same as the tunnel workers. At the end of the shaft stairway, the reporter "plunged directly into the blackness of the rude tunnels … many of them traversed by railways, over which the excavated stone was hauled in car-loads to the hoisting shaft, and all of them were wet with the constant percolation of river water through seams in the roof and walls."[7]

The reporter had descended into a world of tunnels dug at right angles to other tunnels. These galleries were supported by more than 160 pillars, called "piers." Moving forward, "holding their flaring lamps above their heads and splashing through pools of black water," the reporter noted the "regular chuff, chuff, chuff, of the drills, and the sharp rattle of their steel points against the hard rock" as the holes continued to be made for the cartridges.

His tour completed, the writer noticed on the trip back to Astoria a large sign on Big Mill Rock, just across the reef from Flood Rock, that read, "Nitroglycerine—Don't Land."

Varied publications noted that the general "gave directions and had experiments made on the forces to be used to explode by sympathy, and sent two officers to Florida where the water is warmer than [Hell Gate]. It may not be generally known that when the water is cold dynamite may not explode." Experiments that Newton labeled "elaborate" had been going on through General Abbot at Willets Points at the Corps of Engineers School. And Newton's explosive needed to be in liquid form "because we cannot use it in the granular state on account of the weight necessary to have in each drill-hole."[8]

As the testing went on, Newton had to keep money—the costs of the explosives—in mind during this time of funding denied. He had been tasked with the destruction of Flood Rock and the work, in fits and starts, had been going on for more than ten years. To some degree, Newton had insufficient information for many years of the tunneling. The nine acres measured only the surface of the rock, not 60 feet down.

A chart printed in 1881 showed the power of 19 different compounds based on their percent of nitroglycerine. The chart measured "Rendrock" with three different percentages of nitroglycerine and this compound was soon re-named "Rackarock." Rackarock had the advantage of giving off fewer sickening fumes than dynamite. "Rackarock, moreover, is an inert explosive, and whilst costing little more than half as much as dynamite, it possesses somewhat greater efficiency underwater."[9] Rackarock was one of Dr. Hermann Sprengel's potassium chlorate explosives; he could claim that his mixture of 21 percent nitrobenzene had superior destructive ability—108 percent of dynamite. The compound Rackarock was similar to nitroglycerine. A thorough study by General Humphreys of the power of Rackarock stated that with "dynamite No. 1 being taken as standard at 100, then Nitroglycerine (scientific name 'trinitro-glycerin') was accorded an 81 and Rackarock at 104."[10]

General Abbot cautiously concluded that

> Rackarock possesses the merits of high intensity of action, unusual density, absolute safety in handling and storage (components unmixed), and little cost; on the other hand, under the conditions of my tests, an exceptionally strong detonating primer is essential to develop its full power. Experiment alone can determine whether this defect be equally marked when the charges are confined in drill holes in solid rock.[11]

A nine page test result dated May 25, 1885, signed by Lieutenant Derby read "a charge of Rackarock loaded substantially as the charges in the mine at Hell Gate will be, can be exploded with certainty to a distance of 12 feet (the length of the longest drill hole)."[12] The power of Rackarock was rated 23 points higher than nitroglycerine.

Rackarock, a liquid, came from a factory in Tom's River in New Jersey. Oddly enough, the factory was named the Rendrock Powder Company. The two names seem to be used interchangeably. In Newton's scheme, the 30-grain fulminate cap was to be inserted closer to the date of the detonation. Delivered to one area of Mill Rock was one of Rackarock's ingredients, chlorate of potash, 79 percent of the mixture. To a separate area was delivered di-nitrobenzol 21 percent of Rackarock. Being harmless separately, the two could be stored in large quantities, and conveyed, without danger, to the mixing area.

## Preparing the Cartridges

The mixing-house had been set up and there 50-pound batches of the solid and the black oily liquid were mixed, with the solid wooden hoe, in a lead-lined trough. The great value of this kind of explosive, being as it is composed of a solid and a liquid, consists in the ease and safety with which the liquid mixture becomes absorbed by the solid.

A reporter explained the process: "It looked like a wet and inferior grade of brown sugar as it was being mixed in a leaden trough by two men, who used wooden hoes. Only fifty pounds were mixed at a time, and then it was put into the copper cartridge cases, rammed down with a wooden rammer, and the tops of the cartridges were soldered in place using steam-heated soldering irons."[13] They were then removed as fast as they were filled.

The number of pounds of Rackarock put into drill-holes was 240,399 and of dynamite, 42,331. General Newton was getting ready to explode 282,730 pounds of high explosives in the middle of the East River about 700 feet from Manhattan. The Rackarock would be pushed into the more than 10,000 drill-holes in the roof and the pillars. The whole amount of rock to be broken by the final blast was 270,717 cubic yards, covering an area of about nine acres. The day approached when New York City might witness the destruction of 730,935,900 pounds of rock.

Since a decade has passed while drilling Flood Rock, the surveyors had a clear picture of how much explosive would do the job. Newton and his staff had then calculated where to place each explosive and, more importantly, how much was needed to destroy the nine acre reef. As to the destruction of the rock itself, *Science* magazine of January 8, 1886, later correctly decided the explosives were "equivalent in all to about one hundred and fifty tons of dynamite."[14]

Of Newton's staff, we learn that these three officers—Lieut. Col. Walter McFarland, Lieut. George McC. Derby, and Lieut. Willard, of the Corps of Engineers worked closely with the general both on the Scow and at Flood

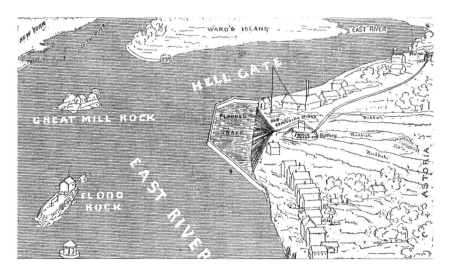

This picturesque map which appeared on the day of the Hallet's Reef's demolition clearly delineates the positions of both Hallet's Reef and Flood Rock. The small tower at the very bottom is the tip of the lighthouse (still in place) at the northern end of Blackwell's (Roosevelt) Island. The map gives an approximation of the three acres of Hallet's Reef and its work area.

Rock. Newton knew that before the project was completed it "will probably cost half a million of dollars and take two years to remove the masses of broken rock." But "when the work is completed the channel will be 1,200 ft. wide, instead of 600, as at present, and 26 ft. deep, enabling ocean steamers to enter at all tides."[15]

As the time for the explosion drew nearer, newspapers were reporting some amazing figures: that the galleries averaged 10 feet by 10 feet in section, but varied in places from 4 to 33 feet in height. The roof averaged 18.8 feet thick. The total length of tunnels was four miles. Supported by 467 stone pillars, (sometimes labeled "piers") each 15 feet square, the mining of the galleries or tunnels had, in effect, hollowed out the giant rock.

A month after the cartridges were being made, this newspaper report appeared:

> Nitroglycerine Explosion. Friday, August 28, 1885—Pittsburg, Penn., August 27—Bradford, Penn., special to the *Dispatch* says: An explosion of 3,000 pounds of nitroglycerine at Roberts Son's factory this morning completely wrecked the building and shook houses for miles around. Two employees were thrown some distance but miraculously escaped with a few slight bruises. The loss will reach several thousand dollars.[16]

All of these accidents let readers know how dangerous these high explosives could be in the wrong hands.

## *Installing the Cartridges*

Twelve of the partitioned boxes of cartridges were taken on board the tug and carried over to Flood Rock, where, with the utmost care, they were landed and lowered down the shaft by elevator into the mine. There were 44,175 of the 24 inch cartridges, all handled with extreme care. Given the caution with which the cartridges were made, transported, brought down, and made ready to insert, the time to complete the task—more than two months—seems understandable.

The cartridge hole-filling continued around mid–August 1885 and would last about two months. Newton's assistants began to reckon how much wire—eventually half a million feet—would be needed to connect the charges in the drilled holes to the electrical source—the batteries on shore. The men who handled the cartridges knew the terrible consequences that might follow a clumsy misstep. London's *The Pall Mall Budget* described how "The workers, twenty picked men, were assigned the job of placing the cartridges…. The ominous silence was unbroken save by the rush, splash, or steady drip of the encroaching waters, and the throb of the great pumps that worked unceasingly to the very end to keep the mine from being flooded."[17]

The workers almost looked to be dressed as fishermen except for the kettle-like lamp worn on the rigid-looking headgear that seemingly protected against both falling debris and leaking water. The twenty men picked for the job were apparently a disciplined lot. So meticulous had Newton's design been that every drill hole was selected and numbered and so those twenty men knew the exact power of the charge for every single drilled hole.

With the bulky clothing, they had to climb ladders or platforms and fill the holes with the cartridges, "pushing them home with wooden rammers. A few inches of the last cartridge were allowed to project beyond the hole," so that after the primers were put in place the sets of wires could be connected.[18] Held in position by little brass spurs projecting from their rims, more than 50,000 of these cartridges were thus placed before the mine was pronounced wholly charged and ready to be exploded. Some charges were tied to the timbers supporting the roof.

The entire explosion had been devised around the experimentation and testing under the guidance of Newton, who instructed Colonel McFarland and Lieutenant Derby of the Engineers to make careful investigation of sympathetic explosions, an idea proposed by Julius H. Striedinger in his 1877 paper "On the Simultaneous Ignition of Thousands of Mines and the Most Advantageous Grouping of Fuses."[19]

McFarland and Derby applied what they knew directly to the unique

Flood Rock explosion, and had demonstrated that one primary blast could set off all the cartridges in a 12-foot hole and others without wires in a 27-foot radius. Using this scheme, it became a matter of arithmetic. How many feet of tunnels needed to be blown up? Place the primary charged holes within 27 feet of each other. Newton was convinced.

With all the holes stuffed full of cartridges something was needed to detonate them. Just as at the explosion at Hallet's, 12 ounces of dynamite in brass tubes served as a priming charge. To set off each primary charge was a "fuse" made of 20 grains of fulminate of mercury. Into the mercury were inserted two wires to complete the circuit. "The inner ends of the wires are united by a small platinum wire. The ends of the wires are then surrounded with fulminate, and the two parts of the tube are put together, that containing the wires slipping within the other. The entire fuse is then covered with gutta-percha."[20] To make more certain the unwired cartridges would explode, the last cartridge in every case was inserted so that its end extended about six inches, so that it might receive the full effect of the shock.

So, rather than wiring all of the holes, there were 591 of these primary charges arranged in 21 circuits of 25 each (525), and three circuits of 22 each (66), all coming together at the poles of each of the 24 batteries, each of the 24 to power each of the circuits. Some of these circuits were nearly a mile long. With the wires attached in the tunnels and routed 300 yards to Astoria, and with the filling of the holes completed, the charge consisted of 141 tons of explosives, distributed over the whole nine acres of the reef. All that remained was the closing of the circuit.

The cartridges were in place by October 3 and what tasks remained involved the transfer of machinery and buildings from Flood Rock to a scow by a force of about 80 men, the scow to be towed to the Astoria side.

## The Guests and the Police

As the scheduled day and time for the explosion was being determined, notices, arrangements, and invitations needed to be planned. The demolition of so large a quantity of rock within the boundaries of this great city brought with it a kind of mix of curiosity, delight, scientific interest and, for some, a way to make some money. Public figures would have to be notified and with those notifications measures needed to be tended to for lodging, transportation to the site, as well as dining arrangements.

Meetings with the New York City Police established placements of officers in Astoria and on Blackwell's Island, and with an extra contingent positioned near the battery house from where the secondary jolt of electricity would flow. Opportunities for participation in the event were quickly requested by civil engineers, scientists, and academics, who might want to take measurements of various kinds, and in different and distant places. One kind of measurement would be photographic evidence of the blast, electronically connected to the initial surge of energy. To reassure the public of the blast's safety, journalists had been given tours and provided with guides of the works. But then came the time to focus on the firing point near the ferry landing at the foot of Astoria Boulevard, the northernmost point in Hallet's Cove.

Alongside the international representatives from Russia, France, Austria, Brazil, and Mexico, stood G.S. Greene, Jr., engineer-in-chief of the Department of Docks, members, General G.S. Greene, president of the American Society of Civil Engineers, and some of the group's members.

"No one had ever attempted an underwater blast on the scale that Newton had proposed for both Hallet's and Flood Rock," wrote historian Marion J. Klawonn in 1977. "Certainly no one had ever tried anything like it in the heart of a large city."[21] Central Park, eventually mid–Manhattan, used gunpowder for most of the blasting of 843 acres, not nitroglycerine or dynamite.

## Explosions Explained

Many of the guests who would witness the explosion from ships on the East River understood in scientific terms very clearly what was about to happen. The chemistry of the detonation was well known ever since the power let loose by the specific combination of chemicals had been documented not long after Nobel's patented products. It was the physics of which the public had but a dim understanding that frightened many of the citizens. Two aspects of an explosion must be understood: pressure and temperature. As for pressure, for example, if you blow too much air into a balloon it explodes. The material strength of the balloon cannot withstand the pressure of the air blown into it.

As in Newton's scheme, if you put a nitroglycerine compound in a confined space, you limit it, like a drilled hole in a rock tunnel. Therefore, its power, the pressure, is enormously strengthened. Flood rock was now dotted with 12,561 holes into which have been jammed 14,000 explosives, themselves confined in wrappings. All 14,000 when ignited simultaneously will create an irresistible pressure on the rocks wherever the explosives have been placed. This pressure is a gas that can expand indeterminately and at great speed. The speed produces the heat. The detonation causes the molecules of explosives to become pressurized gas molecules that move rapidly. Energy—the nitroglycerine—makes pressurized gas. The gas expands.

Once more like a balloon: if you inflate a balloon and just let the air out of it there is no explosion, but if you tie off the balloon—confine the gas—and then prick it with a needle, its pop noisily releases the gas inside. However, in addition, if you light a match under a balloon, it too will blast apart.

This pop, this heated shock wave of nitroglycerin, will move at twice the speed of sound and thus with enormous pressure. In the confined space of the holes holding the explosives, the force on the Rackarock and dynamite would create a supersonic wave heated to a temperature of about 9,000 degrees Fahrenheit. The blast wave itself moved with 1,200 times the original volume of ordinary room pressure and caused the resulting detonation wave to travel at more than 17,000 miles per hour.

To understand the enormity of the destructive power about to be unleashed, consider the detonation at Blossom Rock in San Francisco which broke up 45,825 square feet; Hallet's Reef measured three acres or 130,680 square feet. But Flood Rock was nine acres or 392,040 square feet.

Lieutenant Derby further explained a difference in approaching the problem of Flood Rock. "It was the old idea to undermine and then pulling the prop out from under the bottom and let it fall and lie there." But, Derby continued, while underscoring General Newton's fearless adaption of the new, particularly the new that saved the American taxpayer money, technology had changed. "With the improvement of dredging apparatus, we find it cheaper to do just as little tunneling as will suffice to thoroughly distribute the explosives. The explosives break up the rock into pieces that can be easily handled."[22]

Here it can be understood how important General Newton was to this two decade project. By Newton's rapid implementation of the newer and more efficient technology, his persistence, his precision in planning, in his choice of subordinates like Striedinger, the total cost of the Hallet's Point work was $81,092, and that of Flood Rock, although a three times larger blast, would add up to $106,509, just 24 percent more.

# FOURTEEN

# Newton's Violent Triumph

## *Saturday Night Preparations Before the Sunday Blast*

During the night before the explosion, from 9:30 p.m. until 3:30 a.m., water was siphoned into the shafts to flood the tunnels and galleries. The water served at least three purposes: it would both mute the sound, it would eliminate almost all odors, and it would increase the efficiency of the explosion, as it did at Hallet's. Thus, the 10,000,000 cubic yards of water, working strongly in favor of a controlled explosion, were providing a powerful tamping effect. "All of the vibratory power at Flood Rock was upward," said *The American Naturalist*.[1] How much "tamping"? The weight of the water, given the reliability of the journal above, equaled 16,855,500,000 pounds. This means that when each cartridge explodes, the force is directed even deeper into the rock. When questioned about the effects of the blast, Newton replied that he expects the crowds to see "a huge mass fifteen hundred feet long [rising] in the air perhaps two hundred feet."[2] In place as well were bomb containment blankets. Like the water they "mitigate blasts and eliminate the risk of secondary fragmentation."[3]

Newton's Blast Day was set for October 10, 1885. Four years and four months of actual work time (minus when there was no money available) was marked by caution and thoroughness, two qualities continued on the day of the destruction. *The Brooklyn Daily Eagle* understood the importance of the day and would rush out an edition at 4 p.m. the day of the blast. The paper was published in the third largest city in the country (until it merged with other New York City boroughs 13 years after Flood Rock). The editor of the paper sent reporters to cover the story and they took positions at Astoria, Yorkville, the Stagg Street Armory in the Bushwick section of Brooklyn, and on Blackwell's Island. Shorter pieces from other locations were filed too.

Most of the other New York City papers used approximately the same approach and they too would publish their accounts in the afternoon of the morning explosion.

## Sunday Morning's Tasks

At 5:00 a.m., workers and staff made tests on the wires. Coated in gutta-percha, the electric lines stretched under the water across the river to the ferry landing at Astoria in Queens below Hallet's point in a cove. Some of the circuits were nearly a mile long. At the bomb-proof building, the batteries were connected in such a way that if any few circuits failed through any fault in the connections, an explosion of those charges would still be ensured through the sympathetic action of the neighboring charges. There were 591 primary charges.

By 5:30 a.m., launches identified by a red flag flown in their bows were on station blocking three entrances to the Flood Rock area. Police vessels halted all ships from Long Island Sound, from the piers on both sides of Manhattan and in Brooklyn, and those crafts using the Harlem River.

By 6 a.m. those who saw a profit in explosions were making ready. "In every little cove were flotillas of row boats … lying in wait for the end to gather in the luckless tomcods stunned by the blast," reported the *Daily Eagle* in a later edition on October 10. Solidly in place by 7:30 was a police battalion of 125 formed to protect the area around the firing point in Astoria. General Abbot took charge and made sure that no spectators would get near the house near the steamboat landing. From Willet's Point further north on the river, the site of the United States School of Submarine Mining, officers from the Engineer Corps came to help with security.

Also on duty, by 8:00 a.m., police totaling 100 each in three sections of the Manhattan riverfront were on guard in the streets from 79th to 110th, especially since reports stated that 100,000 might arrive by East Side elevated rail and another 80,000 by horse car. A squad of 50 police was sent to Ward's Island.

In that same eight o'clock hour, city employees spent a busy morning on Blackwell's (Roosevelt) Island at the various buildings of the Department of Charities and Correction located on the island. Many patients and criminals—perhaps 7,000—needed to be taken out of the buildings in case of a powerful shock from the explosion. From the Charity Hospital, 900 patients were wrapped in blankets and taken to cots set up on the lawn. From the penitentiary, 1,096 male prisoners were closely guarded at the upper part of the island supervised by Warden John M. Fox of the Blackwell's Island Workhouse. "The almshouses, including the incurable wards, emptied of the lame, the sick and the blind," wrote the *Times* of October 11. Police squads of 50 each were sent to the upper and lower ends of the Island, partly to keep watch over "1,600 female lunatics." (The city acquired the island in 1828, but the name remained Blackwell's Island while the city operated a prison, "a lunatic asylum, a charity hospital, a smallpox hospital, and a workhouse.")

By 9:00 a.m. crowds begin to form beginning at the area on the shore between 85th and 97th Street, a space called at the time the Harlem Flats. (At a time when uptown was 34th Street, this upper town area was known to be a dumping ground for garbage and as a grazing area for geese and sheep.) The mass soon added up to about 5,000 persons. That space, now called Carl Schurz Park, formed a kind of amphitheater, sloping upwards from the river. Those viewers could easily watch a police steamer patrolling the river as well.

At ten o'clock all commercial navigation through Hell Gate was stopped, and after that time only the swift-moving official launches darted over its waters. For good reason: "The most powerful force which man had ever concentrated," wrote the Brooklyn newspaper, "was about to be loosed."[4]

On the Astoria shore that morning, 10,000 curious New Yorkers circulated, but the early comers began to jostle for the best view; soon another 40,000 would join them. Before long, a crowd of 50,00 people "occupied … summits of lumber piles and houses to the rear extensions of lager beer saloons."[5] People in houses along the shore had been told that if they don't leave it will be at their own risk. Cameramen began their preparations to make timed images of the blast for scientific study.

Now another crowd of watchers, people who had come to see the great event, filed onto the Brooklyn Bridge, more than five miles (100 blocks) distant from Flood Rock. The press of people was so great that it was said that "passing them was a matter of difficulty." Curiously, as the *Eagle* reported, "it was evident … that the burst of water could not be seen because of the turn in the river, yet all stayed hoping to hear a report or feel the shock attendant upon the disintegration of so much rock."[6] Some expressed a fear that the bridge would fall "and several ladies became alarmed but stuck to what they thought their perilous post womanfully."[7]

Watchers on the north side of the narrow Bronx Kill at the edge of the Bronx arrived on foot, by horseback, in public and private carriages and hacks. Spread out from the middle of the shoreline opposite the explosion people stood near the Astoria Ferry Hotel in Queens, but were pushed back from the shoreline by a line of police. Others climbed into trees, to the highest floor of a brewery, or stood on the elevated train platform. Still more were seen by the *Brooklyn Daily Eagle* reporter climbing "to the rear of 'lager beer saloons which backed out over the river.'" Some younger viewers sought out the tops of chimneys or shinnied up trees and lampposts.

Already brought to the firing point was a battery consisting of 35 zinc cells filled with a one to ten mixture of sulphuric acid and water. The cells on this battery would instantly communicate with the 800 cells of the 24 batteries that awaited the signal from the firing point. Four cameras whose shutters were operated via electricity were set up to take images ten seconds apart; the first using the same current as the current sent to the explosives. The Yale University observatory had decided to take a measurement of the blast. The academic scientists had set up an instrument on a New Haven, Connecticut, pier to record the speed with which the tremor from the explosion might reach from the East River to New Haven, 70 miles north and east. From Yale, a Mr. L. Walso stood ready to take the exact time of the detonation.

General Henry Abbot, the prominent military engineer who had contributed to the Corps of Engineers on such topics as Ordnance and Fortifications, and who had worked with Newton on these East River projects, had sent a party of six men and an officer to several points to measure the blast's speed of shock and its power too. Each measuring device was connected by electric wire to the current that exploded the mine. Nearer to the blast, devices had also been placed on Staten Island and nearby Ward's Island.

As the hour approached, at about 10:40, General Newton boarded a tug out to Flood Rock for a final inspection One reporter, years later, did not understand the delay in setting off the detonation.[8] Newton replied, "the holes that been drilled in the tunnel for receiving the nitroglycerine would stretch a distance of over 22 miles If they were placed end to end, and in these holes I am going to put almost 300,000 pounds of explosive which will be discharged simultaneously by means of an electric spark."

The steamer *Castelton,* a ferryboat, was chartered so that about 500 persons including wives and families of the engineers on the project as well as people of status might have a close look at the blast. These passengers had been sent tickets and instructed to meet at the foot of East 33rd Street at ten in the morning of the blast. Generals Hancock, Slocum, and Sheridan (then in command of the United States Army) met on board along with C.C. Martin, who became Washington Roebling's chief assistant on the Brooklyn Bridge, along with a congressman, two of General Newton's sons, as well as several members of the American Society of Civil Engineers. That boat from 33rd Street took its passengers to the Harlem River's output at the lower end of Ward's Island, about two miles from the blast. The little steamer *Chester A. Arthur,* floating next to the *Castelton,* had on board Henry Ward Beecher, a preacher who accepted Darwin. Other crafts—the steamers *Sylvan Dell* and *Sylvan Stream*—"hugged the Harlem shore," along with many other crafts. Policing tugs and launches darted about.

The working tug *General Humphreys* remained on site, as did its tender, to take off the last remaining men from Flood Rock, men who had been ordered to remove the last of the salvageable items from the spot, including the American flag,

> at which time a cheer arose. All of the machinery and most of the buildings had been removed from Flood Rock.... Nothing remained on Flood Rock Island as even a little hut was pulled down.... The last to leave was Lieutenant George McClellan Derby, who, under General Newton, has superintended the entire work, and whose labor of years was now to be crowned with success, or to result in failure.[9]

With three blasts of her whistle as a parting salute, the *Humphreys* moved down the river to a point on the Astoria side just above the ferry slip, where the firing point was. Another sound arose—that of the alarm bell in the Catholic Protectory on Blackwell's "and the children fled out into the spacious yards."[10] On the Manhattan side, the engineers on the steam locomotives pulling the cars of the Third Avenue and Second Avenue elevated lines near 89th Street slowed their trains as much as they thought prudent as the time for explosion approached.

One estimate counted 200,000 watching and waiting. The *Daily Eagle* wrote, "very merry these people were, laughing, smoking and drinking beer and watching ... the busy little tugboats rushing about ... and the great excursions steamers floating lazily past."[11] By now viewers could count many and varied vessels which had come with passengers to watch. These included steamers, yachts, rowboats. Many of those boats sought a way to make money from the event, either before or after the blast.

To demonstrate how fascinating Newton's explosion was thought, even the railroads on the river as a carfloat wanted to be there. "A tug," the *Times* reporter saw, "pulling two flatboats laden with freight cars" halted to wait for the explosion, perhaps out of safety, or perhaps out of curiosity.

As the tenth hour faded, the eyes of 50,000 spectators were focused on the doomed island rock. While Newton's tug sailed its last trip from Flood Rock, warning flags were waved to warn the crowd and ships. From high banks, housetops, fences, steamers, and ice barges, a hundred cameras were directed toward the same point. Soldiers and policemen, some with fixed bayonets, were brought more alerts and many photographers checked their equipment once more.

## Blast Time

An encyclopedia later took note of "a little building which stands on the dock just north of the Astoria Ferry slip and is estimated to be about 1,000 feet from Flood Rock." This is the firing point. Newton's a "huge mass" (1,500 feet long, 200 feet in the air) would appear even though "[t]he resistance offered to the explosives equaled 500,000 tons of rock and 200,000 tons of water."[12]

People on the shore and atop buildings may have come for some sort of show, but few had any idea of the power of the forces that General Newton was about to release. Of the explosion *Science* explained that the strength of the detonation will be expressed "downward, upward and horizontally." To demolish the nine-acre rock, Newton had designed, in effect, a massive, hot hammer moving almost four times faster than the fastest speed of a modern jet aircraft. In the watchers' minds was the *Eagle*'s sentence "the most powerful force which man had ever concentrated was about to be loosened."

With the time for the explosion approaching, the batteries were filled and the 800 plates cranked into the slots by Mr. Striedinger and his small staff. When the electrical circuit is closed, the heat caused by resistance in the electric wire will vaporize the platinum wire which has a melting point of 1,772 degrees Celsius (3,221 degrees Fahrenheit). That temperature produces 588,000 pounds per square inch of pressure on the mercury fulminate primer.

Taken to the landing place at Astoria on the steam yacht *Runaway*, were General Newton, his wife, his daughter Mary Anne, a girl of 11. Also on the steamer were Professor Mercur from West Point, Newton's chief assistant for many years on the East River projects, some military officers, a preacher, and two officers' wives. General Abbot would take charge of the electric key, Striedinger had pushed a wooden block between the contacts of the key that will send the current to Flood Rock.

Brooklyn and New York City papers sent many reporters to different parts of the city. *Harper's Weekly* was not going to be left out and sent the prolific writer Kirk Munroe to cover the event. "Every eye was strained to catch a glimpse of the wonderful spectacle," Munroe wrote. "Fifty thousand faces were turned in the same direction, and fifty thousand hearts beat quick with excitement. An almost deathlike silence settled over the whole scene … and the ripple of the little waves that danced merrily beside the wharves and boats was distinctly audible."[13]

What was not at the scene? No banners, no bands, no ceremonies. No one made a speech, no one praised either the achievement or the designer. It appears Newton did not ask for, nor want, nor need the applause. He needed his work to be done.

## Newton Arrives at the Firing Point

All the watchers could still see the "thousand eddies caused by that massive rock in the East River opposite East 92nd Street in Manhattan.

Seeing the party arrive, some decided to stop their ears with their fingers, certain the blast would sound soon. The key was near a heavy bichromate battery. From there the electric current moved over a single wire beneath the river to a heavy bichromate battery on the rock, which diffused the current through twenty-four independent circuits."[14] Sixty thousand cartridges of high explosives were about to explode simultaneously.

As the time approached for the blast, watchers on Ward's Island "began to seek soft places on the grass where if overthrown by the concussion, they might fall and not get hurt."[15] Flags were run up on the various buildings on Blackwell's Island and a hook and ladder company stood at the ready.

The primary key stood on an old refrigerator in the wooden firing point on the shore 40 yards northeast of the Astoria Ferry slip and 300 yards southwest of Flood Rock. All around this firing point were 450 men and officers from the Corps of Engineers school at Willet's Point standing guard. The method of the connection between the key and the batteries was a set of two wires shielded in rubber which ran from the key to "an inch cable containing seven copper wires." This cable was held in place by being wrapped twice around the dock's post.

Any boats near Flood Rock had been warned by shrieking whistles to depart and sailed away at 11:10. The *Humphreys* blew three loud blasts, a final alert.

General Abbot brought a telegraph key to young Mary Anne and told her to tap it when ready. Abbot then carefully attached the wires to the key. Those still at the firing point moved back and "shrank together" except for Abbot, Newton and his daughter. "Then ensued two minutes of the most intense suspense.... An almost deathlike silence settled over the whole scene, hardly a voice was heard among all the vast throngs of people, and the ripple of the little waves that danced merrily beside the wharves and boats was distinctly audible."[16]

"Little Mary Newton, with her flowing hair," *The New York Herald* reported, "lifts her fair, white hand and places her dainty finger on the key." "Now," says General Newton quietly. "The child's white finger presses the key" at 11:12, setting off "the greatest single discharge of explosive material that has ever been produced by human agency."[17]

The paper went on to remind readers that the blast "was the culmination of nine years' unremitting and scientifically directed labor beneath the waters of Hell Gate, and the finishing touch was given by the hand of the same young lady whose baby fingers, nine years ago, sent a similar electric signal beneath the same waters into the mine at Hallet's Point."[18]

Depending on where a spectator stood, "a rumble might be felt, or not, but certainly a noise could be heard that began like a hiss, or like a growl, and then up shot a column of water, of spray, and some said it rose

*At 11:13:50.2 AM, on October 10, 1885, with all charges in place, the cavern at Flood Rock flooded with water —12 year old Miss Mary Newton pressed the key that set the charge and snapped the shutter of the camera that took this picture.*

**The United States Corps of Engineers provided this photograph of some Astoria residents watching the Flood Rock explosion from roof tops (https://www.nan. usace.army.milPortals37docshistoryhellgate.pdf).**

more than 125 feet into the air." The *Daily Eagle* reporter also noted "Suddenly a tremor was felt and a great while pile rose like a gigantic iceberg ... 200 feet high ... and 200 yards long.... Not a black speck appeared in this mass ... and it seemed like some great palace of alabaster." This great body of water was "immediately followed by great black clouds.... It was no more an island and a ragged stone heap rested in its stead."[19]

The next day, the *New York Times* devoted its entire front page to the blast:

> A deep rumble, then a dull boom, like the smothered bursting of a hundred mighty guns far away beyond the blue horizon, rolled across the yellow river. Up, up, and still up into the frightened air soared a great, ghastly, writhing wall of white and silver and gray. Fifty gigantic geysers, linked together by shivering, twisting masses of spray soared upward, their shining pinnacles, with dome-like summits, looming like shattered floods of molten silver against the azure sky [October 11, 1885].

When the "spray descended, it was followed by a cloud of brownish and yellow gas that floated toward Astoria and quickly dispersed" wrote the *Herald*. A British reporter thought smoky vapor to be "immense black clouds.... No loud report or great shock occurred; the damage was confined to the breaking of a few panes of glass, the falling of some loose ceilings, and the shaking off" of some loose bricks from a chimney at Astoria.[20]

## *After the Blast*

The whole area of the reef was shattered. The dredging of Flood Rock would have to commence. The change in the channel would have to be measured by the Lighthouse Department as that task was now their responsibility. Newton had not only demolished Flood Rock and in such an orderly and secure way that "The effect upon the island and its several structures was not at all what had been expected. Much of the island still remained above-water, though but little raised from its surface." The explosion was so artfully done that it looked as if the entire nine acres had been neatly piled in Hell Gate, awaiting the dredgers. Even "the great wooden [water] tank was somewhat shattered, but still stood much as it had before the explosion."[21]

*Engineering News and American Contract Journal*, a trade publication of the period, judged, "Although the volume of explosives used here exceeded six-fold the greatest charge ever previously fired in the world, the work of the engineers was so well done and the precautionary measures of Gen. Newton so well taken, that no accident or delay of any kind occurred."[22]

The rocks shown on the far left are probably the west end of the Middle Reef (in top photograph, opposite page).

The explosion over, "The crowd, with a sudden impulse, broke and ran to be clubbed back by the police" described the *Daily Eagle*. "Looking out at what had been the nine acres of Flood Rock, the large derrick that had hoisted so many hundreds of cars of debris, could be seen on its side and those bulkheads that had buttressed the Island were gone."

"The fisher boats now dashed out," continued the *Daily Eagle*, "still uncertain if danger lay ahead, but eager for the easy hauling up of dead and stunned fish." "Small boats swarmed upon the river like crows in a cornfield," wrote the *Times*, calling the persons wishing to be taken to the remains of Flood Rock "relic hunters." Some men in rowboats headed for the rock itself and the smoldering wooden ruins, excited to be so near to the blast and yet safe. Hundreds paid twenty-five cents to be rowed out to Flood Rock so that the next day many might say "I walked on Flood Rock." Others, scooping their hands into the river, held up some stone or a scorched piece of timber. "It was said that from the detonation until darkness there was not a boat for hire from 84th Street in Manhattan to about 104th Street, at the Harlem River Bridge. At one point in the afternoon,

The viewers may be standing on the Manhattan side at East 92nd Street, from atop George Ehret's brewery (*Frank Leslie's Illustrated Newspaper*, October 22, 1885).

A drawing includes a view of the skeletal tower holding a light at Hallet's Point (*Harper's Weekly*, October 17, 1885).

about 200 stood on the shattered pile of rock. Toward dusk, some took the timber and made rafts from it."[23]

"Everything worked like a charm," the *Brooklyn Daily Eagle* determined in the story on its front page just hours after the detonation. On the islands near Flood Rock, "at noon there was nothing to show out of the ordinary routine work of the institutions had occurred."[24]

General Newton, observed to have a cut on his neck afterward, realized it was from a wire attached to one of the cameras and said: "I am the only one injured. ... Everything worked like a charm. There was need for neither firemen or physician ... nor was there any disturbances among prisoners or inmates in any of the institutions.... At noon there was nothing to show that anything out of the ordinary routine work of the institutions had occurred."[25]

And "Immediately after the explosion the government steamer *John Rodgers* began to take soundings above the reef ... and within half an hour after Miss Mary Newton touched the electric button, Lieutenant Debry was found directing the movements of the dredging scows, which were set to work immediately to pick up and clear away the vast quantity of debris, in the shape of broken rock, that he has deposited at the bottom of the river. This material will be taken down to the great 100-foot hole at the head of Blackwell's Island and there dumped overboard. It is expected that the dredging process will be continued for about three years."[26]

One commentator called it "Newton's scientific violence." There is not one other example of a planned explosion like Flood Rock. No one since Newton has attempted a peacetime explosion like Flood Rock. No one since has ever used Rackarock and dynamite in such large quantities, nor used dynamite for a civil engineering project. No one since Newton has ever attempted an underwater demolition project of this size. In the opinion of one of the engineering journals, "Although the volume of explosives here used exceeded six-fold the greatest charge ever previously fired in the world, the work of the engineers was so well done and the precautionary measures of General Newton so well taken, that no accident or delay of any kind occurred."[27]

The blasted rock continued to shift and settle. The Turntable, a whirlpool formerly near Flood Rock, shifted 200 feet away, due to the destruction of the rock that had caused the whirlpool in the first place. The new whirlpool that had the power to hold boats almost stationary was moved nearer to Astoria's shore.

The *New York Times*, and other papers, devoted its entire front page to the event and newspapers all over the world from Chicago to Australia reported on the explosion of the nine-acre boulder. "Ships and boats ... which had never before been allowed through Hell Gate, were now able to navigate this passage easily. Shipping trade increased to some $4 million worth of cargo a day, justifying the millions of dollars for the un-snagging operation," Marian Betancourt asserted. "A new life was infused into the Port of New York, which was once again able to assert itself as the undisputed leading port in the nation at the time."[28]

In honor of Newton's achievement, Flags were run up every flagstaff on Blackwell's Island, by order of the Commissioner of Charities and Corrections Brannan. Even the next day, many people chose to board the 92nd Street ferry just to take a look at the explosion site. The *Times* estimates 5,000 came to visit.

## Estimates of Work to Be Done

After the Flood Rock detonation, in the judgment of *Popular Science Monthly,*

> All the problems … of that stupendous work were completely and conscientiously studied out; and the accuracy of the studies was fully … realized in every part of the labors. The resources of science were drawn upon with an unerring vision of their scope; the appliances of engineering art were employed with precise adaptation to their purpose, and an exact measurement of the effect they were intended to produce; so that in all that has been achieved there has been no failure and no waste.[29]

Back in 1869, $44.28 per cubic yard was the amount needed by Maillefert. The total cost of the work done on Flood Rock in 1885, including the final blast, amounted to $2.99 per cubic yard of the whole amount of rock broken, or 85.66 percent less than the cost of breaking Hallet's Point.

Dredging, of course, needed to commence. The general was aware that this labor of many years had a great result since it "serve[s] … the commerce of New York and greatly adds to its volume." He estimates that another $300,000 will be needed to complete the dredging and that on hand the operation has but $130,000 yet to be expended.

It is not known how much of Flood Rock lay below the tunnels and piers. Nor has it ever been said that the explosion shattered what was above the 65 foot deep tunnels as well as what might have been below. On an NOAA map consulted in 2020, the deepest part of the East River in what might have been the position of Flood Rock has been surveyed at 85 feet.

A journal in Britain estimated "It will probably take at least a couple of years more to move the debris when a splendid channel will be conferred upon the European steamship traffic to New York. This operation has certainly been a great triumph for American engineering skill."[30] The *New York Times* labeled it "another triumph of human skill over the resistance of nature." In 2008, Philip Lopate concluded, "Hell Gate was tamed, ending one of the most heroic, arduous endeavors in the city's history."[31] No one for many years would dare a planned explosion of such magnitude.

## Some Measurements of the Explosion

"Assuming the figures of the Cambridge report as correct, and that the mine at Flood rock was exploded at 11:14, seventy-fifth meridian time, it took the wave just 194 seconds to travel 190 miles, or at the rate of 5,120 feet per second."[32]

Newton's contributions to science included the shock wave's power that, once known, might be the source of further study. In ordinary ways it was learned a few plates fell off shelves and shattered on Blackwell's Island, near to the explosion. The shock was felt at Columbia College, in Central Park, at Vassar, Yale, Harvard, in Washington, D.C., and in Bayshore and Patchogue, Long Island, and at West Point where Newton's assistants had been sent to take measurements. Patchogue, 48 miles away, registered the shock reaching that point 21½ seconds after detonation.

More startling, a slight effect was observed at Cambridge, Massachusetts, where vibrations were recorded at Harvard University, 185 miles distant, 200 seconds after the explosion, indicating a rate of transmission of the shock wave at 4,900 feet per second.

Newton had helped change an explosion into an event for study. Yale University observatory recorded the blast and would now figure out how fast a wave motion travels through the earth's crust.

# FIFTEEN

# Newton's Last Decade

Newton could lay claim to working not only on Hallet's, Flood Rock, and other rocks in mid-river, but also on the Hudson River, the Harlem River. His surveys continued hundreds of miles from New York Harbor, up the Hudson, and into Lake Champlain in Vermont. He even engaged on that other entrance to the harbor in his "Report Upon the Work of Deepening Gedney's Channel Through Sandy Hook Bar, New York," published on January 12, 1885.

Fewer than two years after the Flood Rock triumph, Newton had reached the Army's mandatory retirement age of 64 and was officially separated from the service on August 27, 1886. That same year, he received the Laetare Medal from the University of Notre Dame, awarded to a practicing Catholic in honor of service to society. The fourth recipient of the medal, he remains the only engineer to receive the award in this honor's 130-plus year history.

Leaving his post in D.C., he returned to his home at 40 West 75th Street, a half block from Central Park. He left behind him not only a 19-year job well done, to say the total cost was about 86 percent less than the cost of breaking Hallet's Point.[1] At about the same time as his retirement, it was written of him,

> The resources of science were drawn upon with an unerring vision of their scope; the appliances of engineering art were employed with precise adaptation to their purpose, and an exact measurement of the effect they were intended to produce; so that in all that has been achieved there has been no failure and no waste ... it would be hard to produce a higher testimonial to the value of the faithful pursuit of the studies ... than these achievements at Hell-Gate.[2]

New York City mayor William Grace, recognizing his superior skill, probity, and ability to work with his staff, appointed him Commissioner of Public Works on August 28, 1886. *The New York Times* that day thought Newton's "appointment will be followed by ... a complete deliverance of political intrigues." The fact that Grace was notable for being the first Roman Catholic mayor of New York City may have convinced him to employ the general given Newton's high standing among American Catholics.

And among engineers, he was as highly regarded for his science as he was for his strict control of public monies. The *Engineering News Record* in October of 1886 remembered that Newton was

> long identified with the promotion of the commercial interests of New York City ... and ... the duties of this office are most important [for] the general improvement of the city, and ... General Newton is eminently fitted by his ability as an engineer, experience as an executive officer, and undoubted probity of character, to fill the office in a manner creditable to his own reputation, and profitable to the citizens of New York."[3]

Later it was written that "from the day of this appointment until the expiration of his term, he conducted that great city department in a way that it had never been conducted before," in the opinion of the *Engineering and Mining Journal* of May 4, 1885, "Politics were absolutely eliminated from the department, all sorts of reforms were instituted, and an ironclad rule was made that there should be no removals [firings] save for good and sufficient cause."

Before his term expired on November 24, 1888, he accepted the presidency of the Panama Railroad Company on April 2 which position he filled until his death.

## Death

General John Newton died on May 1, 1895, from chronic rheumatism and pneumonia at his home at 40 West 57th Street. He is buried at West Point and, like Abraham Lincoln, Newton had his infant son Francis, who died in 1873, buried with him.

## Honors

Newton's widow and children could read the laudatory comments that appeared in newspapers, magazines and journals. One recalled that the *New York World* had put the Flood Rock explosion on its front page, calling Newton's work "A Momentous Event in the Progress of Science." From the *Engineering and Mining Journal*, "General Newton's work excited the admiration of engineers all over the world. ... General Newton ... carefully and conscientiously studied the problems, and the accuracy of his conclusions was shown in the exact correspondence of results with the object that he sought."[4]

One year after Flood Rock's destruction, *The Popular Science Monthly* wrote:

> he has shown himself a man who held the resources of science in his hand, and knew exactly what to do with them; and in the use he made of them, to promote the greatest public benefit, the originality of the devices which he contrived, and the certainty with which he accomplished his designs, he has shown himself to possess the highest title to scientific recognition.[5]

John Newton as a civilian.

Upon his death, his accomplishments other than Hell Gate were summed up this way:

> General Newton was superintending engineer of the construction of the defences on the Long Island side of the entrance to New York harbor; of the improvements of the Hudson River, and of the fort at Sandy Hook; the improvement of the Harlem River; of the Hudson River between Troy and this city; of the channel between New Jersey and Staten Island, and of a number of harbors on Lake Champlain.[6]

From the National Academy of Sciences on May 3, 1895, came this appreciation:

> His preeminence as an engineer who had successfully met every problem confronting him … was fully recognized at home and abroad…. We give grateful acknowledgement to the rare fidelity, the patient and studious devotion, the far seeing discernment, and the unwavering adherence to principles which, with a ripe judgment made his counsels so valued and his administration so successful. … [His] invariable amiability and courtliness made business associations with him a rare pleasure.[7]

Of course, some projects started by Newton were not completed until after his death. One such was the Harlem River Shipping Canal:

**Tim Clifford's sculpture "Monument to a Missing Island" commemorates the destruction of Flood Rock (courtesy Tim Clifford).**

This canal adds nearly fifteen miles of water-front, the real Source of New York's commercial supremacy, to the city's equipment, makes a safe and easy passageway from the East River to the Hudson, gives increased business advantages to the "annexed district," and marks the partial completion of a work which the late General John Newton, famous for his skill at Hell Gate in improving navigation, planned originally.[8]

After his death a steam tug named *General Newton* worked on the Delaware River and was still in service in 1912. In 1899 the Corps of Engineers named a new 175-foot paddle wheeler in Newton's honor. During her 58 years of service the *Gen. John Newton* functioned as a freight, mail, and passenger packet as well as a patrol boat. She provided rescue service during major floods by transporting emergency rations and medical supplies to stricken areas and evacuating persons stranded by rising water.

It was used as a maritime courthouse and visited by several presidents, and in 1958, the Engineers sold it to the University of Minnesota to use as showboat for the University of Minnesota's annual summer productions. It lasted a hundred years until it was destroyed by fire in 2000.

The drill scow *General Newton*, of his design and named after him, remained in service at least until 1906. The steam tug *General Newton* was still at work on the Hudson at least until 1911.

The sculptor Tim Clifford made this monument to Newton's work and it stands on the banks of the East River.

When it came time for Marian Betancourt to select those who would be included in her 2016 book *Heroes of New York Harbor: Tales from the City's Port*, the choice for her Chapter Five was General John Newton "The Man Who Opened Hell Gate."

# A Gazetteer of the
# East River Obstructions

Bald-headed Billy—fifteen yards from Hatter's dock at the southern bend of Hallet's Point just north of Hallet's Cove.

Two Barents Islands—Later Great and Little Barn islands and then named Ward's and Randall's Islands.

Great and Little Barn islands—now joined as Randall's Island.

Bayard house—110th Street or 82 Jane Street.

Belmont Island and U-Thant Island—Hazard-to-navigation later covered over with tunnel-building rubble. Mid-river off East 42nd Street.

Benson's Point—known later as Rhinelander's Point.

Blackwell's Island—1872 lighthouse in service—see Roosevelt Island.

Blackwell's Rock—still unknown.

Bounty's dock or wharf—in Pot Cove (and see Pot Rock).

Bread and Cheese—off northern tip of Blackwell's.

Battery Reef—still unknown.

Bronx Kill—separates Randall's Island from Port Morris, a narrow passage that extends westward from the East River to the Harlem River. A fixed railroad bridge with a clearance of 68 feet and a fixed highway bridge with a clearance of 51 feet cross the passage. NOAA says Bronx Kill is navigable but not recommended as a route of travel. It is shoal and obstructed throughout.

Bushwick Inlet—the area around First Street in Williamsburg (present-day Kent Avenue) with Franklin Street in Greenpoint.

Charlotte Rock—opposite 34th Street, northwest of entrance to Newtown Creek.

Coenties Reef—about 200 yards from Pier 8, mainly loose rock, 250 feet long, with a maximum width of 130 feet. Originally an artificial inlet in the East River, it was often used for the loading and unloading of ships and was land-filled in 1835.

Coenties Slip—a boat slip between the wharves of South Street. Its triangular shape was determined by the curve of the shoreline: two sides radiating from the apex at Pearl Street into a wide mouth of water. Whitehall Street, with its counting houses, was part of the back office.

Corlears Hook—located at the intersection of Jackson and Cherry streets along the Franklin Delano Roosevelt (FDR) Drive on Manhattan's Lower East Side—at 397 FDR Drive or opposite 10th Street in Brooklyn.

Diamond Reef—between Governor's Island and the Battery—a bed of rock, 366 feet long and 255 feet wide.

Ferry Reef—34th Street—250 by 100 feet; 7 feet deep—not touched until 1889.

Flood Rock—nine acres in mid-channel, or 392,040 square feet, so large that a powerful whirlpool called The Turntable formed around it; 1,200 feet long by 602 feet wide, but only 250 square feet width showed. Flood Rock lies about 300 yards from the light on Hallet's Point. The least water over it is 18 feet. Located about 1,000 feet north-easterly from Hallet's Point, at Astoria.

Frying Pan Rock (or Shoal)—halfway between Hallet's Point and Ward's Island, 200 feet by 100 feet; originally 12,475 sq feet.

Gibbs's point, in Queens, Hallet's Cove, Astoria, situated between Ravenswood Houses at 21st Avenue and 34th Street, Astoria.

Governors Island—the older name was Nutten's Island.

Graves Point—Vessels going from the East River into the channel leading to the Harlem River, on the northwest side of the island, must pass up round Graves Point; 2,407 feet from tip of Bronx to Manhattan's 95th Street.

The Gridiron—about 500 feet into Flood Rock—the Gridiron was a connected part.

Hallet's Cove—where Vernon Boulevard terminates at the intersection of east-bound Main Avenue and Eighth Street. Ferry landing was at Fulton Street, now Astoria Boulevard, Astoria.

Hallet's Point—Hallet's Point is southward of Ward's Island. A light is shown and fog bell rung from a small, white pyramidal wooden tower on the northern end of the point.

Hancock Rock—just off Astoria.

Harlem Cove—now Manhattanville.

Harlem Kills—also called Bronx Kills—once 480 feet wide, 24 feet deep; then 100 yards wide and 4–10 feet deep in 1874; in 1913, 80 feet wide at bridge.

Harris Hook—foot of East 86th Street.

Hatter's Dock—the northern point Between Ward's Island and Ringgold's Dock, near Bald-Headed Billy, the northern point of entrance to Hallet's Cove.

Heel Tap Rock—off south end of Roosevelt Island—23,000 square feet; 53,500 cubic yards.

Hen and Chickens Reef—with Negro Head, making up the Middle Reef, possible to include Little Negro Head—off 91st Street.

High Bridge—between Manhattan and Bronx, earliest bridge, 1848, as part of the Croton Aqueduct system; its walkway was completed in 1864.

Hog's Back Rock—900 feet from Hallet's point near Ward's Island—off south end of Randall's Island—appears joined with Holmes Rock.

Holmes Rock—off south end of Randall's Island, just south of Hog Back. "Holmes Rock and Hog Back are two bare rocks, which are on the eastern and northern parts, respectively, of a reef in the bight on the south side of Ward's Island westward of Negro Point. The western extremity of this reef is marked by a light" (NOAA).

Horn's Hook—89th–90th Streets and East End Avenue.

Hoyt's place—700 yards east of Hallet's Point.

Hoyt's Rock—perhaps named for The Hoyt mansion, located in Astoria between 24th and 25th avenues.

Lawrence Point—northwest tip of Astoria—may be older name for Willet's Point.

Little Dock Street—now named Water Street.

Little Hell Gate—strait was 1,100 yards in length, 165–200 yards wide, 8–24 feet deep. It split Randall's and Ward's islands. What remains is opposite 112th St. in Manhattan and is a constructed salt marsh or inlet with a boardwalk path and arch bridge for pedestrians and bikers. Informational signs and decorative fencing indicate that the site is a nature preserve. "Little Hell Gate, which formerly separated Ward's Island from Randalls Island and formed a passage from East River to Harlem River, has been mostly filled in and together with Sunken Meadow joins Ward's Island with Randalls Island" (NOAA).

Man-O-War Rock/Reef—later Belmont Island and U-Thant Island.

Middle Ground—area between Randall's Island and upper Astoria. Narrowest section measures 905 feet.

Middle Reef—Flood Rock proper and the reefs known as the Negro Heads, Hen and Chickens, and Gridiron.

Mill Rock—1,000 feet off the eastern edge of 96th Street: Little and Great Mill Rocks, an interval from Flood Rock of about 800 feet. Also called Leland Island and occupied by Sandy Gibson, they were originally two smaller islands, named Great Mill Rock and Little Mill Rock. In 1890, rock fill from the blast was used to close the gap between the two islands. Now listed at 3.996 acres. The NOAA description is "Mill Rock, on the northwestern side of the main channel through Hell Gate, is 0.2 mile southwest of Ward's Island and the same distance northwest of Hallet's Point. The islet is marked by lights on its north and south ends."

Morris dock—same as Port Morris, which borders the Bronx Kill, East 132nd Street.

The Narrows—narrowest part, 1.01 miles wide.

Negro Head—with Hen and Chickens, making up the Middle Reef, possible to include Negro Head and Little Negro Head—off 91st Street.

Negro Point—southeastern point of Ward's Island, a place frequented by black fishermen at the time. "Negro Point is the southernmost point of Ward's Island. Triborough Bridge, which crosses East River from Negro Point to Long Island 6.8 miles from The Battery, has a highway suspension span with a clearance of 138 feet" (NOAA).

Nes Rock—off 19th Street.

Newtown Creek—begins near the intersection of 47th Street and Grand Avenue on the Brooklyn-Queens border at the intersection of the East Branch and English Kills. It empties into the East River at 2nd Street and 54th Avenue in Long Island City, opposite Bellevue Hospital in Manhattan at 26th Street.

North Brother Island—2,000 feet northwest of Riker's, owned by Edward Ackerson in the early 1800s, has about 13–20 acres of land. Distance to South Brother Island was 850 feet. Lighthouse beacon light activated first time in November 1869. Lighthouse keepers lived on the island. Home to Riverside Hospital beginning in 1885. After World War II, the military built homes there for returning veterans. After more than a half-century sitting abandoned and neglected, the lighthouse has collapsed, and now barely a glimpse of its roof can be seen above the foliage that has grown up on the island. The city bought it in 2007.

Nutten's Island—see Governor's Island.

Pilgrim Rock—19th Street; discovered 1884 when a passenger steamer named *Pilgrim* struck it.

Polhemus Dock—Long Island at intersection of River Road and Wolcott Avenue, now 21st Avenue.

Pot Cove—around the bend of Hallet's Point; its shore lies 600 feet from the current Ways Reef.

Pot Rock—just outside of Pot Cove between Ward's Island and North Astoria: 12,125 square feet; 33,500 cubic yards. This obstacle contributed to the formation of a whirlpool called "The Pot."

Quinters Reef—still unknown.

Randall's Island—was known as Minnahanonck, Little Barn Island, Buchanan's Island and Montresor's Island, with Ward's, 480 acres.

Ringgold's Dock—near Ward's Island.

Rock off 3rd Street—not named.

Rock off 26th Street—not named.

Roosevelt Island—former names include Varcken's, Hog, York, Manning's, and Blackwell's; 88.5 acres, 668 feet from northern tip with its lighthouse to Astoria. Now called Welfare Island.

Rylander Point or Reef—off 92nd Street.

Satterlee's Dock—at 220th street.

Scaly Rock—still unknown.

Shelldrake Rock—in Pot Cove.

Shelldrake Reef, or "lying close inshore at Hallet's Cove; combined Way's Reef and Shelldrake; 6,200 square feet.

Shell Reef—off 9th Street.

Sherman's Creek—Tenth Ave. between Academy St. and the Harlem River, south of 201st Street.

Sound River—very early name for the East River.

South Brother Island—6 acres, extends 1,300 feet to Riker's Island; space between is only 18 feet deep.

Sunken Meadow—land once separate from Randall's Island but now part of it.

Throggs Neck—When John Throckmorton arrived in 1642, it was called Maxson's point, but within a few decades the neck of land was identified as "Frockes Neck." No longer private by 1836, the area served as a kind of summer resort if only as a day trip by steamer from the south.

U Thant Island—see Man-O-War Rock and Belmont Island.

Van Kenlen's Hook—at the southern bank and outlet of the Harlem River.

Ward's Island—255 acres, parallel to 99th to 115th streets—Tekenas, Great Barn Island. Bodies were moved from potter's field in Bryant Park in 1840, and in 1857 bodies moved from East Side between 48th and 50th streets—1,790 feet to Hallet's Point. Area was designated a graveyard for the poor and remained so until 1840, when thousands of bodies were moved to Ward's Island.

Washington Parade Ground—now Washington Square Park, used as potter's field cemeteries until bodies moved to 48th to 50th and Fourth and Lexington. In 1857 bodies again moved to 70 acres on Ward's Island.

Way's Reef—235 feet long by 115 feet wide near Woolsey's Bath House near Pot Cove; 6,200 square feet; 44,200 cubic yards. Way's Reef was surveyed at 235 feet by 195 feet.

Wallabout Bay—between the present Williamsburg and Manhattan bridges, opposite Corlear's Hook on Manhattan to the west, across the East River. Wallabout Bay is now the site of the Brooklyn Navy Yard.

Woolsey's Bath House—north of Hell Gate and around the bend of Hallet's Point past Pot Cove, there is a Woolsey Street there; AKA Woolsey's Stone Bath House.

Washington Parade Ground—(later called Reservoir Square) now Washington Square Park, used as potter's field cemeteries until bodies moved to 48th–50th and Fourth and Lexington. In 1857 bodies moved to 70 acres on Ward's island. Now site of the New York Public Library main branch.

# Chapter Notes

## PREFACE

1. "Drain the Oceans: Detonating Flood Rock." https://www.nationalgeographic.com.au/videos/drain-the-oceans/drain-the-oceans-detonating-flood-rock-5971.aspx.

## INTRODUCTION

1. "Obituary," *Engineering and Mining Journal*, May 4, 1895, 419.

## CHAPTER ONE

1. Ryan Healy, "The Strange History of NYC's Mighty Hell Gate," February 22, 2016 1:42 p.m. https://gothamist.com/news/the-strange-history-of-nycs-mighty-hell-gaten.
2. "Transportation Developments in the Early Republic." https://www.connerprairie.org/educate/indiana-history/travel-and-transportation, n.p.
3. James Buckingham, *America, Historical, Statistic, and Descriptive*, Volume 2 (Fisher, Son & Co., 1841), 481.
4. John Harrison Morrison, *History of New York Ship Yards* (New York: Wm. F. Sametz, 1909), 54.
5. James Kent, *A Practical Treatise on Commercial and Maritime Law. With a chapter on Incorporeal Hereditaments, etc.* (London: Thomas Clark, 1837), 171.
6. Ric Burns et al., *New York: An Illustrated History* (New York: Alfred A. Knopf, 2003), 141.
7. Edward Williams, "Statistics of the City of New York." *The New York Annual Register* (New York: J. Leavitt, 1832), 225.
8. Robert Curley, *The Complete History of Ships and Boats* Britannica Educational Publishing. *The Complete History of Ships and Boats: From Sails and Oars to Nuclear-Powered Vessels* (Rosen Publishing Group, 2011), 144.
9. John Harrison Morrison, *History of New York Ship Yards*, 19.
10. John Leander Bishop et al., *A History of American Manufactures from 1608 to 1860* (United Kingdom, E. Young, 1868), 127.
11. Robert Curley, *The Complete History of Ships*, 19.
12. Charles Ingersoll, *History of the Second War Between the United States of America and Great Britain*, Volume 1 (Philadelphia, Lippincott, Granbo and Co. 1852), 42.
13. Klein Daniel B. and John Majewski. "Turnpikes and Toll Roads in Nineteenth-Century America." eh.net/encyclopedia/turnpikes-and-toll-roads-in-nineteenth-century-america/.

## CHAPTER TWO

1. Art. 4. The Blasting of Rocks Underwater Without Drilling." *Merchants' Magazine and Commercial Review*. Volume 27, 1852, 322.
2. *Ibid.*
3. *Ibid.*, 323.
4. *Journal of the Senate of the State of New York*. S. Southwick 1850, 172.
5. John Newton. "East River and Hell Gate Improvements." *The Popular Science Monthly*. February 1886, 446.
6. "Navigating the East River." *Blue Seas*. https://www.offshoreblue.com/cruising/east-river.php.
7. *Report on the Tides and Currents of Hell Gate: One of the Entrances to New York Harbor Made to the Superintendent of the*

*Coast Survey.* (Washington, DC: U.S. Government Printing Office, 1869), 4.

8. The Great Submarine Blast at Hallet's Point." *The Friend.* 7 October 1876: 61.

9. Cyrus B. Comstock, "Memoir of John Newton 1823–1895." National Academy of Sciences, 13 November 1901 237.

10. *Ibid.*

11. Ingersoll, Charles Jared. *History of the Second War Between the United States of America and Great Britain, Declared by Act of Congress, the 18th of June, 1812, and Concluded by Peace, the 15th of February, 1815.* (Lippincott, Grambo & Co., 1852), 42.

12. Wheeler, David Everett. *The New York Harbor and the Improvements Necessary for Its Accommodation of Commerce and the Removal of the Dangers at Hell Gate: a Paper ... Before American Geographical and Statistical Society.* (New York: J.F. Trow, 1856), 12.

13. *New York Times*, November 18, 1851, 2.

14. Approaches to New York." *Harper's.* July 1884, 278.

15. "Remembering the Fall River Line." https://blog.mass.gov/transportation/mbta/remembering-the-fall-river-line/.

16. *Documents of the Assembly of the State of New York* (New York: E. Croswell, 1852), 59.

17. "The Explosion." *Brooklyn Daily Eagle.* October 10, 1885, 1.

18. John Newton. "East River and Hell Gate Improvements." *The Popular Science Monthly,* February 1886, 443.

19. "Blasting Rock Under Water Without Drilling." *The Merchants' Magazine and Commercial Review.* September 1852, 323.

20. Newton. "East River," 436.

21. *Proceedings—Proceedings of the Board of Councilmen of the City of New York* (New York: Edward Jones and Co., 1862), 24.

## Chapter Three

1. *New York Times* July 23, 1865, 3.

2. The Unbarring of Hell Gate." *Scribner's Monthly.* November 1871, 37.

3. *Ibid.*, 38.

4. "The Great Submarine Blast at Hallet's Point." *The Friend.* October 7, 1876, 61.

5. "State Tonnage Duties on National Commerce." *New York Times,* June 8, 1860, 4.

6. Stephen Sears. *Chancellorsville* (New York: Houghton, Mifflin Harcourt, 2014), I.

7. "Sketch of General John Newton." *The Popular Science Monthly.* October 1886, 836.

8. *New York Herald,* April 18, 1866, 6.

9. "Nitro-Glycerine as Freight" "Nitro-Glycerine as Freight—An Important Suit and Heavy Damages." *New York Times,* August 22, 1867, 2.

10. *United States Congressional Serial Set* (Washington, DC: U.S. Government Printing Office, 1867), 2.

11. "Hell Gate—The Growth of the City." *New York Herald,* 10 September 1876, 3.

12. William Lette Stone. *History of New York City from the Discovery to the Present Day* (New York: E. Cleave, 1868), 244.

13. "An Elevated Railroad." *New York Tribune,* January 25, 1866, 4.

14. Perkins, Sid. "When Horses Really Walked on Water." *The Chronicle of the Horse.* May 21, 1999, 91.

15. Newton "East River and Hell Gate Improvements," 439.

16. "The Great Submarine Blast," 61.

17. *Annual Report of the Secretary of War* (Washington, DC: U.S. Government Printing Office, 1869), 734.

18. Sanger, George P. *Public Laws of the United States of America. Carefully Collated with the Originals at Washington* (New York: Little, Brown & Company, 1867), 422.

19. "The Great Submarine Blast," 62.

20. *United States Congressional Serial Set,* 1867, 10.

21. *Ibid.*, 7.

22. *Ibid.*

23. "Improvements," 436.

24. Newton. "East River and Hell Gate," 436.

25. *Annual Report of the Secretary of War,* 1869, 737.

26. American Experience. "The Transcontinental Railroad Nitroglycerin." www.pbs.org/wgbh/americanexperience/features/tcrr-nitroglycerin/.

## Chapter Four

1. *Annual Report of the Secretary of War* (Washington, DC: U.S. Government Printing Office, 1867), 75.

2. *Report of the Superintendent of the Coast Survey* (Washington, DC: U.S. Government Printing Office), 159.

3. Davenport, M. *Under the Gridiron. A Summer in the United States and the Far West, including a Run through Canada* (Canada: Tinsley Brothers, 1876), 23.

4. Newton. "East River and Hell Gate Improvements," 441.

5. *Report on the Tides and Currents of Hell Gate: One of the Entrances to New York Harbor Made to the Superintendent of the Coast Survey* (Washington, DC: U.S. Government Printing Office, 1869), 3.

6. *Report of the Chief of Engineers* (Washington, DC: U.S. Government Printing Office, 1869), 739.

7. Stephen R. Brown. *A Most Damnable Invention: Dynamite, Nitrates, and the Making of the Modern World* (New York: Macmillan, 2005), 96.

8. American Experience. "The Transcontinental Railroad Nitroglycerin." www.pbs.org/wgbh/americanexperience/features/tcrr-nitroglycerin/.

9. "East River Improvement Association." *New York Herald*, 11 December 1868, 4.

10. *Report of the Chief of Engineers*, 1869, 424.

11. "Hell-Gate, in the Channel of the East River." *Potter's American Monthly*. December 1875, 894.

12. Stone, William Lette. *History of New York City from the Discovery to the Present Day* (New York: E. Cleave, 1868), 244.

13. *Ibid.*, 29.

14. Newton, "East River and Hell Gate Improvements," 439.

15. *Report of the Chief of Engineers*, 1869, 727.

16. *New York Times*, August 22, 1867, 1.

17. *Titusville Morning Herald*, July 11, 1868, 1.

18. *New York Times*, October 1, 1868, 2.

19. Abbot, Charles Greeley, et al. *Biographical Memoir of Henry Larcom Abbot, 1831–1927*. National Academy of Sciences, 1930, 7.

20. *New York Herald*. January 16, 1869, 4.

21. Williams, J. David. *America Illustrated*. De Wolfe, Fiske & Company, 1883, 2003, 63.

22. *Report of the Chief of Engineers*, 1870, 434.

23. "Hell Gate: M. Maillefert's Blasting—His Work to Be Surveyed by the Government—Hallett's Point." *New York Times*, November 30, 1869, 5.

24. *New-York Tribune*, January 12, 1869, 3.

25. "Nitroglycerine Its Use and Its Danger." *Scientific American*, February 1, 1868, 73.

26. *Annual Report of the Secretary of War* (Washington, DC: U.S. Government Printing Office, 1870), 392.

27. "Hallett's Point Reef. Progress."

28. Barnard, 869.

29. *The Military Engineer*, 337.

30. "Mining Operations," 887.

31. *Report of the Chief of Engineers* (Washington, DC: U.S. Government Printing Office, 1870), 432.

32. "Hell Gate: M. Maillefert's Blasting," 5.

33. *Ibid.*

34. *Report of the Chief of Engineers*, 1870, 432.

## CHAPTER FIVE

1. *Report of the Chief of Engineers* (Washington, DC: U.S. Government Printing Office, 1886), 681.

2. *Annual Report of the Secretary of War* (Washington, DC: U.S. Government Printing Office, 1870), 434.

3. Annual Report of the Secretary of War (Washington, DC: U.S. Government Printing Office, 1871), 927.

4. "The Ocean Steamer." *Harper's New Monthly Magazine*. No. CCXLII—July 1870—Vol. XLI, 186.

5. *Economic Trends of War and Reconstruction, 1860–1870* (New York: Brookmire Economic Service, 1918), 27.

6. Byron, Carl. *A Pinprick of Light: The Troy & Greenfield Railroad and Its Hoosac Tunnel.* (Hartford: New England Press, 1995), 34.

7. "A Frightful Nitroglyercine Explosion." *The Plantation*. Plantation Publishing Company, April 9, 1870, 182.

8. "Inaugural Address." *Transactions of the Society of Engineers*. Published by the Society, 1887, 123.

9. Knox, Thomas W. *Underground, or Life Below the Surface: Incidents and Accidents Beyond the Light of Day; Startling Adventures in All Parts of the World; Mines and the Mode of Working Them; Under-Currents of Society; Gambling and Its Horrors; Caverns and Their Mysteries; The Dark Ways of Wickedness.* (Hartford: J.B. Burr, 1878), 89.

10. *Ibid.*, 68.

11. "Unbarring," 36.

## CHAPTER SIX

1. "The United States Drilling Scow, East River." *Scientific American.* Munn & Company, July 19, 1879, 38.
2. *Ibid.*, 39.
3. John Newton. *Report of the Operations of the U.S. Steam-Drilling-Scow in the Removal of Way's Reef* (Washington, DC: U.S. Government Printing Office, 1875), 6.
4. *War Department Exhibit: Corps of Engineers, Catalogue* (Washington, DC: U.S. Government Printing Office, 1893), 12.
5. "Unbarring," 53.
6. *Ibid.*
7. "Unbarring," 51.
8. "Department of Docks Excursion."
9. Schermerhorn, 194.
10. Ryan, n.p.
11. Lobell, n.p.

## CHAPTER SEVEN

1. "Art. 4. The Blasting," 324.
2. Newton. "East River and Hell Gate Improvements," 439.
3. "American Industries, No. 2," 49.
4. Newton. "Report of the Operations," 3.
5. *Reports of Practice*, 30.
6. "The Government Works," 40.
7. *Annual Report of the Secretary of War*, 1873, 943.
8. *Manufacturer and Builder Magazine*, 153.
9. *War Department Exhibit*, 9.

## CHAPTER EIGHT

1. *Report of the Chief of Engineers*, 1875, 804.
2. "Unbarring," 53.
3. *Ibid.*
4. "Unbarring," 51.
5. "Unbarring," 241.
6. Ralph Whitney. "The Unlucky Collins Line." *American Heritage*, February 1957, n.p. https://www.americanheritage.com/unlucky-collins-line.
7. Wingate, Charles Frederick, and Henry C. Meyer. *The Engineering Record, Building Record and Sanitary Engineer*, August 25, 1900, 177.
8. *Report of the Chief of Engineers (War Department Exhibit: Corps of Engineers, Catalogue* (Washington, DC: U.S. Government Printing Office, 1893), 209.

9. *Ibid.*
10. *Ibid.*
11. Ellis, 361.
12. *Annual Reports of the War Department* (Washington, DC: U.S. Government Printing Office, 1875), 204.
13. Newton. *Report of the Operations*, 10.
14. *Annual Reports of the War Department*, 1875, 208.
15. *Official Gazette of the United States Patent Office.* U.S. Department of Commerce. *Department* (Washington, DC: U.S. Government Printing Office, 1874), 759.
16. Newton. *Report of the Operations*, 6.
17. *Ibid.*
18. Stevens, John Austin. "Progress of New York in a Century, 1776–1876. An Address Delivered Before the New York Historical Society, December 7, 1875." New York Historical Society (New York: Printed for the Society, 1876), 48.
19. Newton. *Report on the Operations*, 9.
20. *Annual Report of the War Department* (Washington, DC: U.S. Government Printing Office, 1875), 247.
21. *Report of the Chief of Engineers.* (Washington, DC: U.S. Government Printing Office, 1875), 208.
22. "Deep Water."
23. "The Explosion." *Brooklyn Daily Eagle*, October 10, 1885, 1.

## CHAPTER NINE

1. "Removal of the Hell Gate Rocks." *Scientific American.* Scientific American, Incorporated, September 30, 1876, 214.
2. Report of the Board on Behalf of United States Executive Departments at the International Exhibition, Philadelphia, PA, 1876 (Washington, DC: U.S. Government Printing Office, 1884), 324.
3. "Removal of Flood Rock, New York City." *Scientific American.* Scientific American, Incorporated, September 24, 188X, 197.
4. D.P. Heap, *The Engineer Department, U.S. Army, at the International Exhibition, 1876* (Washington, DC: U.S. Government Printing Office, 1884), 322.
5. Davenport, *Under the Gridiron*, 21.
6. *Annual Report of the Secretary of War* (Washington, DC: U.S. Government Printing Office, 1875), 686.
7. "Modern Explosives and Their Practical Application." *Engineering News*, July 5, 1884, 2.

8. *"Report of the Chief of Engineers,"* 1875, 204.

9. "Hell Gate Open. Successful Destruction of Hallett's Point Reef. A Triumph of Science." *New York Herald*, September 25, 1876, 1.

10. Stephen Chester. "Nitroglycerine: Its Manufacture and Use." *Transactions of the American Society of Civil Engineers*, June 2, 1869, 124.

11. *Decisions*, 489.

12. "Hell Gate." *Engineering News*. September 30, 1876, 313.

13. *Report of the Chief of Engineers* (Washington, DC: U.S. Government Printing Office, 1908), 441.

14. *Ibid.*

15. *Annual Report of the Secretary of War* (Washington, DC: U.S. Government Printing Office, 1877), 236.

16. "Blown to Death." *New York Herald*, September 5, 1885, 1.

17. "Dynamite Unchained." *Sunday Mercury*, September 24, 1876, 1.

18. "Blowing Up of the Hell Gate Obstructions." *Frank Leslie's Illustrated Newspaper*, October 7, 1876, 71.

19. "The Hell Gate Improvements." *Engineering and Mining Journal*, October 24, 1885, 287.

20. "Hell Gate Explosion." *New York Herald*, September 22, 1876, 5.

21. "Hell Gate Destroyed—The Explosion." *Harpers Weekly*, October 14, 1876, 17.

22. "Hell Gate Explosion," 5.

## CHAPTER TEN

1. "Hell Gate Destroyed—The Explosion." *Harpers Weekly*, October 14, 1876, 17.

2. "Telegraphic News. Hell Gate. The Great Explosion A Complete Success. No Accident of Any Kind." *Daily Albany Argus*, September 25, 1876, 1.

3. Leveson Francis Vernon-Harcourt. *Achievements in Engineering During the Last Half Century* (New York: Charles Scribner's Sons, 1891), 166.

4. "Gen. Newton's Big Blast—The Triumph of Science on the Reefs Under Hell Gate." *The Sun*, September 25, 1876, 1.

5. "The Great Mine Fired." *New York Tribune*, September 25, 1876, 1.

6. "Hell Gate Open. Successful Destruction of Hallett's Point Reef. A Triumph of Science." *New York Herald*, September 25, 1876, 1.

7. "The Great Mine Fired," 1.

8. "Gen. Newton's Big Blast—The Triumph of Science on the Reefs Under Hell Gate," 1.

9. "Hell Gateless." *Rutland Weekly Herald*, September 28, 1876, 2.

10. "Hell Gate Open," 1.

11. "Hell Gate Blown Up." *Sacramento Daily Union*, September 25, 1876, 2.

12. *Ibid.*

13. "Hell Gate: The Scene of the Explosion." *Columbus Daily Enquirer*, September 30, 1876.

14. "Gen. Newton's Big Blast," 1.

15. "Hell Gate Open," 1.

16. "Hell Gate Blown Up," 2.

17. "Gen. Newton's Big Blast," 1.

18. "The Great Mine Fired." *New York Tribune*, September 25, 1876, 4.

19. *Ibid.*

20. "Hell Gateless," 2.

21. "The Hell Gate Explosion." *The Popular Science Monthly*. D. Appleton, November 1876, 105.

22. "The Explosion at Hell Gate." *The Internal Revenue Record and Customs Journal*, October 2, 1876, 313.

23. "The Great Blast." *The Sailors Magazine and Seamen's Friend*. New York: American Seamen's Friend Society, November 1876, 337.

24. "Hell Gate Open," 1.

25. "Mary Newton at the Hell Gate Explosion." *The Pacific Rural Press and California Farmer*, November 4, 1876, 5.

26. "Hell Gate Open," 1.

27. "The Great Blasting." *New York Tribune*, September 26, 1876, 2.

28. "Hell Gate Open," 1.

29. "The Great Blast," 337.

30. "The Great Mine Fired," 4.

31. "Telegraphic News. Hell Gate. The Great Explosion: A Complete Success. No Accident of Any Kind." *Daily Albany Argus*, September 25, 1876, 1.

32. "Hell Gate Open," 1.

33. "Rending Hell-Gate Rocks; The Submarine Mine Exploded." *New York Times*, September 25, 1876, 3.

34. "Hell Gate Open," 1.

35. "Rending," 3.

36. "Newton's Channel. Soundings and Operations at the Site of the Blast. The Reef

Destroyed." *New York Herald*, 26 September 26, 1876, 4.

37. *Annual Reports of the War Department*. (Washington, DC: U.S. Government Printing Office, 1877), 238.

38. "Hallett's Point. The Mine a Complete Success. Two Hundred Feet of Shore Line Swept Away." *Commercial Advertiser*, September 25, 1876, 2.

39. "Newton's Channel," 4.

40. *Report of the Chief of Engineers* (Washington, DC: U.S. Government Printing Office, 1877), 238.

41. "Newton's Channel," 4.

42. *Appletons' Annual Cyclopædia and Register of Important Events Embracing Political, Military, and Ecclesiastical Affairs; Public Documents; Biography, Statistics, Commerce, Finance, Literature, Science, Agriculture, and Mechanical Industry* (New York: D. Appleton & Company, 1888), 382.

43. "Hell Gate Ajar."

44. "The Explosion at Hell Gate," 313.

45. "Newton's Channel."

46. Knox, *Underground*, 897.

47. Farrow, 311.

48. "The Explosion at Hell Gate," 313.

49. "Rending," 3.

50. "Work at Newton's Channel." *New York Times*, December 26, 1876. WPA Writers Project. Maritime History of New York. Going Coastal Inc., 2004.

51. *Annual Report of the Corporation of the Chamber of Commerce, of the State of New York for the Year* (New York: New York Chamber of Commerce Press, 1877), 44.

52. "Mission." National Academy of Sciences. www.nasonline.org/about-nas/mission/.

53. "Preparing to Open Hell Gate." *Harper's Weekly*. Harper's Magazine Company, October 10, 1876, 661.

## CHAPTER ELEVEN

1. "Removal of the Hell Gate Rocks," 214.

2. *Appletons' Annual Cyclopædia and Register of Important Events Embracing Political, Military, and Ecclesiastical Affairs; Public Documents; Biography, Statistics, Commerce, Finance, Literature, Science, Agriculture, and Mechanical Industry* (New York: D. Appleton & Company, 1886), 392.

3. *New York Herald*, October 8, 1876, 5.

4. *Annual Report of the Chief of Engineers*, 1877, 228.

5. "Preparing to Open Hell Gate," 660.

6. *Annual Reports of the Secretary of War*, 1877, 228.

7. "The Life Blood of a Nation Is in Its Commerce." To the United States Congress assembled. Charles Stoughton. Washington, DC. December 1, 1879. www.loc.gov/resource/rbpe.20605200n.p.

## CHAPTER TWELVE

1. "Sketch of General John Newton." *The Popular Science Monthly*, October 1886, 839.

2. Dr. Jean-Paul Rodrigue. "The Emergence of Mechanized Transportation Systems." https://transportgeography.org/?page_id=995.

3. *Marine Engineer and Naval Architect*, November 1, 1885, 197.

4. *Annual Report of the Chief of Engineers* (Washington, DC: U.S. Government Printing Office, 1879), 376.

5. "The Government Works at Hell Gate." *Manufacturer and Builder Magazine*, July 7, 1879, 208.

6. *Report of the Chief of Engineers* (Washington, DC: U.S. Government Printing Office, 1879), 378.

7. "The Government Works," 208.

8. Annual *Report of the Chief of Engineers*, 1879, 376.

9. "The High Explosives in War." *The United States Army and Navy Journal and Gazette of the Regular and Volunteer Forces*. July 5, 1884, 993.

10. Nichols, Theodore A. *The Caribbean Gateway to Colombia: Cartagena, Santa Marta, and Barranquilla and Their Connections with the Interior, 1820–1940*. Berkeley: University of California, 1951, 86.

11. The New American Cyclopaedia: A Popular Dictionary of General Knowledge. Vol XII (New York: D. Appleton & Company, 1869), 288.

12. "The Excavation of Flood Rock, Hell Gate." *Scientific American*. October 16, 1880, 249.

13. Newton. "East River and Hell Gate Improvements," 443.

14. "The Excavation" 249.

15. "Removal of the Hell Gate Rocks." *Scientific American*, September 30, 1876, 215.

16. *Ibid*.

17. Newton. "East River and Hell Gate Improvements," 445.

18. "Preparing to Open Hell Gate," 660.

19. *Report of the Chief of Engineers* (Washington, DC: U.S. Government Printing Office, 1883), 536.

20. *Ibid.*, 537.

21. Congressional Record: Proceedings and Debates of the Congress (Washington, DC: U.S. Government Printing Office, 1884), 4972.

22. *Club Men of New York: Their Occupations, and Business and Home Addresses: Sketches of Each of the Organizations* (College Alumni Associations. Republic Press, 1893), 21.

23. Newton. "East River and Hell Gate Improvements," 270.

24. *Congressional Record: Proceedings,* 95.

25. "Assumption," 130.

26. Jeannette (Edwards) Rattray. *Perils of the Port of New York: Maritime Disasters from Sandy Hook to Execution Rocks* (New York: Dodd, Mead & Company, 1973), 43.

27. *Annual Report of the Secretary of War* (Washington, DC: U.S. Government Printing Office, 1883), 538.

## Chapter Thirteen

1. Sarah Vowell. *Assassination Vacation* (New York: Simon & Schuster, 2005), 127.

2. Newton. "East River and Hell Gate Improvements," 435.

3. "Ready to Blow Up Flood Rock. Expected Effects of the Explosion." *New-York Tribune*, October 1, 1885, 3.

4. Newton. "East River and Hell Gate Improvements," 436.

5. *Report of the Chief of Engineers,* 1883, 535.

6. "Preparing to Open Hell Gate," 660.

7. *Ibid.*, 662.

8. *Congressional Record Proceedings,* 4792.

9. Vernon-Harcourt. "Blasting Operations," 270.

10. "Dynamite in War." *The American Engineer*, April 13, 1887, 133.

11. "General Abbot's Latest Experiments." *Engineering and Mining Journal*, January 12, 1884, 26.

12. *Annual Report of the Secretary of War*, 1885, 724.

13. "Preparing to Open Hell Gate," 662.

14. "General Abbot's Report on the Flood Rock Explosion." *Science.* American Association for the Advancement of Science, January 8, 1886, 25.

15. "Inaugural Address." *Transactions of the Society of Engineers*, 1887, 22.

16. "Nitro Glycerine Explosion." *New York Herald*, August 28, 1885, 3.

17. "Opening Up Hell Gate." *The Pall Mall Budget: Being a Weekly Collection of Articles Printed in the Pall Mall Gazette from Day to Day, with a Summary of News.* London, 22 October 1885, 109.

18. "Preparing to Open Hell Gate," 662.

19. Striedinger, Julius H. "On the Simultaneous Ignition of Thousands of Mines and the Most Advantageous Grouping of Fuses." *Engineering.* September 21, 1877, 236.

20. Newton "East River and Hell Gate Improvements," 447.

21. Marion J. Klawonn. *Cradle of the Corps: A History of the New York District, U.S. Army Corps of Engineers, 1775–1975.* (Department of Defense, Department of the Army, Corps of Engineers, New York District, 1977), 85.

22. "Ready to Blow Up Flood Rock," 3.

## Chapter Fourteen

1. Ballou, 139.

2. *Ibid.*

3. Security Pro USA.

4. "The Explosion at Hell Rock." *Engineering News and American Contract Journal.* October 2, 1886, 221.

5. *Ibid.*

6. "The Shock on the Bridge."

7. "The Explosion at Hell Rock," 221.

8. "Story of the Man."

9. "Blowing Up Flood Rock," 679.

10. "The Explosion at Hell Rock," 221.

11. *Ibid.*

12. Chambers, 215.

13. Munroe, Kirk. "Blowing Up Flood Rock." *Harper's Weekly.* October 17, 1885, 679.

14. *Missionary Review*, 548.

15. "The Explosion at Hell Rock," 221.

16. "Blowing Up Flood Rock," 679.

17. Munroe, "Blowing Up Flood Rock," 679.

18. "At the Electric Key."

19. "The Explosion at Hell Rock," 221.

20. Harcourt, 272.

21. "Blowing up Flood Rock."

22. "The Explosion at Hell Rock," 249.

23. "Rending Hell-Gate Rocks."

24. "The Explosion at Hell Rock," 221.

25. *Ibid.*

26. Munroe, 679.

27. "Editorial Notes." *Shipbuilding & Marine Engineering International*, November 1, 1885, 198.

28. Marian Betancourt. *Heroes of New York Harbor: Tales from the City's Port.* (Guilford, CT: Globe Pequot Press, 2016), 70.

29. "Sketch of General John Newton." *The Popular Science Monthly*. October 1886: 840.

30. *Marine Engineer and Naval Architect*, 198.

31. Philip Lopate. *Waterfront: A Walk Around Manhattan*. New York: Knopf Doubleday, 2008), 338.

32. "The Flood Rock Explosion." *Scientific American*. November 21, 1885, 324.

## CHAPTER FIFTEEN

1. Newton. "East River and Hell Gate Improvements," 447.

2. "Sketch of General John Newton," 840.

3. "The Explosion at Hell Rock." *Engineering News and American Contract Journal*. October 2, 1886, 241.

4. "Obituary," 419.

5. "Sketch of General John Newton," 840.

6. "Obituary," 419.

7. "Annual Reunion" United States Military Academy. Association of Graduates. John Newton, June 10, 1895, 240.

8. "The Harlem Ship Canal," 605.

# Bibliography

Abbot, Charles Greeley, et al. *Biographical Memoir of Henry Larcom Abbot, 1831–1927*. Washington, D.C.: National Academy of Sciences, 1930.

Abbot, Henry L. "Reports, Experiments and Investigations to Develop a System of Submarine Mines, for Defending the Harbors of the United States." *Professional Papers of The Corps of Engineers, U.S. Army. No. 23*. Washington, D.C.: Government Printing Office, 1881.

Abbot, Henry L. "Shock of the Explosion at Hallet's Point." *Annual Reports of the War Department*. Washington, D.C.: United States War Department, 1886.

Abbott, Benjamin Vaughan, and Austin Abbott. *Reports of Practice Cases, Determined in the Courts of the State of New York: With a Digest of All Points of Practice Embraced in the Standard New York Reports*. New York: J. S. Voorhies, 1860.

Albion, Robert Greenhalgh. *Rise of New York Port 1815–1860*. Boston: Northeastern University Press, 1984.

Ambrose, Stephen E. *Nothing Like It in the World: The Men Who Built the Transcontinental Railroad 1863–1869*. New York: Pocket, 2005.

American Experience. "The Transcontinental Railroad Nitroglycerin." www.pbs.org/wgbh/americanexperience/features/tcrr-nitroglycerin/.

*Annual Report of the Chief of Engineers to the Secretary of War for the Year*. Washington, D.C.: U.S. Government Printing Office, 1877.

*Annual Report of the Chief of Engineers to the Secretary of War for the Year*. Washington, D.C.: U.S. Government Printing Office, 1884.

*Annual Report of the Chief of Engineers to the Secretary of War for the Year*. Washington, D.C.: U.S. Government Printing Office, 1889.

*Annual Report of the Corporation of the Chamber of Commerce, of the State of New York, for the Year*. New York: Press of the Chamber of Commerce, 1877.

*Annual Report of the Director, United States Coast and Geodetic Survey, to the Secretary of Commerce*. Washington, D.C.: U.S. Government Printing Office, 1874.

*Annual Report of the Secretary of War*. Washington, D.C.: U.S. Government Printing Office, 1867.

*Annual Report of the Secretary of War*. Washington, D.C.: U.S. Government Printing Office, 1870.

*Annual Report of the Secretary of War*. Washington, D.C.: U.S. Government Printing Office, 1871.

*Annual Report of the Secretary of War*. Washington, D.C.: U.S. Government Printing Office, 1873.

*Annual Report of the Secretary of War*. Washington, D.C.: U.S. Government Printing Office, 1877.

*Annual Report of the Secretary of War*. Washington, D.C.: U.S. Government Printing Office, 1883.

*Annual Report of the Secretary of War*. Washington, D.C.: U.S. Government Printing Office, 1885.

*Annual Report Upon the Improvement of Rivers and Harbors in the Vicinity of New York City, and in Northern New Jersey, in Charge of Walter McFarland: Being Appendix E of the Annual Report*. Washington, D.C.: U.S. Government Printing Office, 1870.

*Annual Reports of the War Department*.

Washington, D.C.: U.S. Government Printing Office, 1875.

*Annual Reports of the War Department.* Washington, D.C.: U.S. Government Printing Office, 1876.

*Appletons' Annual Cyclopaedia and Register of Important Events: Embracing Political, Civil, Military, and Ecclesiastical Affairs; Public Documents; Biography, Statistics, Commerce, Finance, Literature, Science, Agriculture, and Mechanical Industry.* New York: Appleton, 1876.

*Appletons' Annual Cyclopaedia and Register of Important Events Embracing Political, Military, and Ecclesiastical Affairs; Public Documents; Biography, Statistics, Commerce, Finance, Literature, Science, Agriculture, and Mechanical Industry.* New York: Appleton, 1888.

"Art. 4. The Blasting of Rocks Underwater Without Drilling." *Merchants' Magazine and Commercial Review* 27 (1852): pp. 320–329.

Ascher, Kate. *The Works: Anatomy of a City.* London: Penguin Books, 2007.

"Assumption in News Making." *The Phrenological Journal and Life Illustrated,* February 1885.

"At the Electric Key. The Vast Mine Exploded by General Newton's Little Daughter, Mary." *New York Herald,* October 11, 1885.

Ballou, William Hosea. "The Flood Rock Explosion." *The American Naturalist,* February 1886, pp. 137–140.

Barnard, Frederick Augustus Porter. *Johnson's New Universal Cyclopaedia: A Scientific and Popular Treasury of Useful Knowledge.* New York: A. J. Johnson, 1877.

Baynes, T. Spenser. *The Encyclopaedia Britannica.* Edinburgh: Supplement 1884–1889.

Bell, Blake A. "Using a Massive Explosion to Market Pelham Manor Real Estate in 1876." January 30, 2017. http://historicpelham.blogspot.com/2017/01/using-massive-explosion-to-market.html.

Betancourt, Marian. *Heroes of New York Harbor: Tales from the City's Port.* Guilford, CT: Globe Pequot Press, 2016.

Bishop, John Leander, et al. *A History of American Manufactures from 1608 to 1860: Comprising Annals of the Industry of the United States in Machinery, Manufactures and Useful Arts, with a Notice of the Important Inventions, Tariffs, and the Results of Each Decennial Census.* United Kingdom, E. Young, 1868.

Blaskowitz, Charles. *A Plan of the Narrows of Hells-Gate in the East River, Near Which Batteries of Cannon and Mortars Were Erected on Long Island with a View to Take Off the Defenses and Make Breaches in the Rebel Fort on the Opposite Shore to Facilitate a Landing of Troops on New York Island.* [1776] Map. Retrieved from the Library of Congress, www.loc.gov/item/gm71000922/.

"Blasting Rock Under Water Without Drilling." *The Merchants' Magazine and Commercial Review,* September 1852, pp. 320–329.

"Blowing up of the Hell Gate Obstructions." *Frank Leslie's Illustrated Newspaper,* October 7, 1876, p. 71.

"Blown to Death." *New York Herald,* September 5, 1885.

Boland, Ed Jr. "F.Y.I." *New York Times,* February 2, 2003.

"Bomb Blankets." *Security Pro USA.* www.securityprousa.com/collections/bomb-blankets.

Booth, Mary Louise. *History of the City of New York: From Its Earliest Settlement to the Present Time.* New York: W. R. C. Clark & Meeker, 1859.

Brodhead, John Romeyn. *History of the State of New York.* New York: Harper & Brothers, 1853.

Brown, Henry Collins. *Valentine's Manual of Old New York.* New York: Valentine's Manual, 1927.

Brown, Stephen R. *A Most Damnable Invention: Dynamite, Nitrates, and the Making of the Modern World.* New York: St. Martin's Press, 2005.

Buckingham, James. *America, Historical, Statistic, and Descriptive—Volume 2.* London: Fisher, Son & Co., 1841.

Burns, Ric, James Sanders and Lisa Ades. *New York: An Illustrated History.* New York: Knopf, 2003.

Byron, Carl. *A Pinprick of Light: The Troy & Greenfield Railroad and Its Hoosac Tunnel.* Brattleboro, VT: Stephen Greene Press, 1978.

"Canmeyer vs. Newton." *Reports of Patent Causes Decided in the Circuit Courts of the United States Since January 1.* January 20, 1877, pp. 287–291.

*Chambers' Encyclopaedia: A Dictionary of*

*Universal Knowledge.* Edinburgh: W. & R. Chambers, 1892, 215.

Chester, Stephen. "Nitroglycerine: Its Manufacture and Use." *Transactions of the American Society of Civil Engineers,* June 2, 1869.

Chrastina, Paul. "Engineers Detonate Huge Blast in New York." *Commentaries on American Law,* 1876.

"The Clearing of The Hell Gate Obstructions." *Frank Leslie's Popular Monthly,* November 1876.

*Club Men of New York: Their Occupations, and Business and Home Addresses: Sketches of Each of the Organizations: College Alumni Associations.* New York: Republic Press, 1893.

"The Common Carriers." *The Black Diamond,* August 5, 1922, pp. 170–171.

Comstock, Cyrus B. "Memoir of John Newton 1823–1895." Washington, D.C: National Academy of Sciences, 1901.

*Congressional Record: Proceedings and Debates of the Congress.* Washington, D.C.: U.S. Government Printing Office, 1884, p. 4972.

Curley, Robert. *The Complete History of Ships and Boats: From Sails and Oars to Nuclear-Powered Vessels.* New York: Rosen Publishing Group, 2011.

Davenport, Montague. *Under the Gridiron. A Summer in the United States and the Far West, Including a Run Through Canada.* London: Tinsley Brothers, 1876.

"Debate Future of Former Link Between Randalls and Wards Islands." *New York Times,* April 16, 1995.

*Decisions of the Commissioner of Patents and of the United States Courts in Patent and Trademark and Copyright Cases.* Washington, D.C.: U.S. Government Printing Office, 1880.

"Deep Water at Hell Gate." *The Week: A Resumé of Current Opinion, Home and Foreign,* November 16, 1872. p. 667.

Denton, Daniel. *A Brief Description of New York, Formerly Called New Netherlands.* New York: W. Gowans, 1845.

DePalma, Anthony. "East River Fights Bids to Harness Its Currents for Electricity." *New York Times,* August 13, 2007.

"Department of Docks Excursion by the Commissioners." *New York Tribune,* October 18, 1870.

*Documents of the Assembly of the State of New York.* New York: E. Croswell, 1852.

*Documents of the Senate of the State of New York.* New York: E. Croswell, 1856.

Domville-Fife, Charles W. *Submarine Engineering of Today: A Popular Account of the Methods by Which Sunken Ships Are Raised, Docks Built, Rocks Blasted Away, Tunnels Excavated, and Many Other Feats of Engineering Beneath the Surface of the Water.* Philadelphia: J. B. Lippincott, 1914.

"Dynamite in War." *The American Engineer,* April 13, 1887, pp. 129–134.

"Dynamite Unchained." *Sunday Mercury,* September 24, 1876.

"East River Improvement Association." *New York Herald,* December 11, 1868.

*Economic Trends of War and Reconstruction, 1860–1870.* New York: Brookmire Economic Service, 1918.

"Editorial Notes." *Shipbuilding & Marine Engineering International,* November 1, 1885, p. 198–199.

Ehrenman, Gayle. "Digging Deeper in New York." *Mechanical Engineering Magazine,* November 2003. https://asmedigital collection.asme.org/memagazineselect/ article/125/11/51/379537/Digging-Deeper-in-New-York1f-the-Third-largest.

"An Elevated Railroad." *New York Tribune,* January 25, 1866.

Ellis, W. H. "Nitroglycerine: Its History, Manufacture, and Industrial Applications." *The Canadian Journal,* November 1873.

*Encyclopedia of American Industries: Service & Non-Manufacturing Industries.* Farmington Hills, MI: Gale Research, 1995.

"Enterprise on the Water." Maritime Nation, 1800–1850. https://americanhistory.si. edu/onthewater/exhibition/2_3.html.

"The Excavation of Flood Rock, Hell Gate." *Scientific American,* October 16, 1880, p. 249.

"The Explosion." *Brooklyn Daily Eagle,* October 10, 1885.

"The Explosion at Hell Gate." *The Internal Revenue Record and Customs Journal,* October 2, 1876, pp. 313–314.

"The Explosion at Hell Rock." *Engineering News and American Contract Journal,* 1885, pp. 249–250.

Farrow, Edward Samuel. "Submarine Drilling." *Farrow's Military Encyclopedia. A Dictionary of Military Knowledge.* New York: Military-Naval Pub., 1895.

"The Flood Rock Explosion." *Scientific American*, November 21, 1885.

"A Frightful Nitroglycerine Explosion." *The Plantation*, April 9, 1870, p.182.

Gannon, Michael. *Operation Drumbeat: Germany's U-Boat Attacks Along the American Coast in World War II*. New York: HarperCollins, 2011.

"Gen. Newton's Big Blast—The Triumph of Science on the Reefs Under Hell Gate." *The Sun*, September 25, 1876.

"General Abbot's Latest Experiments." *Engineering and Mining Journal*, January 12, 1884. pp. 26–27.

"General Abbot's Report on the Flood Rock Explosion." *Science*, 1886, pp. 25–26.

Ghose, Tia. "Happy Birthday, Dynamite: Interesting Facts About the Explosive Material." *LiveScience*, May 7, 2017. https://www.livescience.com/59000-interesting-facts-about-dynamite.html.

Gilje, Paul A. "On the Waterfront: Maritime Workers in New York City in the Early Republic, 1800–1850." *New York History* 77, no. 4 (October 1996): pp. 395–426.

Goldstein, Harold. "Slips of Old New York." jondreyer.org/hal/slipsofoldnewyork.html.

Goodrich, Frank. *Ocean's Story; or, Triumphs of Thirty Centuries*. Philadelphia: Hubbard Bros. 1875.

Gordon, Robert. *The Rise and Fall of American Growth: The U.S. Standard of Living Since the Civil War*. Princeton, NJ: Princeton University Press, 2017.

Gorokhovich, Yuri. "The Application of GIS to the Analysis of Historical Coastline Changes in New York City." *Proceedings–AAG Middle States Division*, volume 24, 1991. https://msaag.aag.org/wp-content/uploads/2013/04/16_Gorokhovich.pdf.

"The Government Works at Hell Gate." *Manufacturer and Builder Magazine*. July 7, 1879, p. 208.

"The Great Blast." *The Sailors Magazine and Seamen's Friend*, November 1876, pp. 337–338.

"The Great Mine Fired." *New York Tribune*, September 25, 1876.

"The Great Submarine Blast at Hallet's Point." *The Friend*, October 7, 1876, pp. 60–62.

The Greater Astoria Historical Society, Erik Baard, Thomas Jackson, and Richard Melnick. *The East River*. Charleston, SC: Arcadia Publishing, 2005.

Guiterman, Arthur. *Ballads of Old New York*. Ann Arbor, MI: Scholarly Publishing Office, University of Michigan Library, 2005.

"Hallett's Point. The Mine a Complete Success Two Hundred Feet of Shore Line Swept Away." *Commercial Advertiser*, September 25, 1876.

"Hallett's Point Reef. Progress in the Art of Excavating Under Water. Triumphs Over Great Difficulties." *New York Times*, September 16, 1876, p. 3.

Harcourt, Leveson. "Blasting Operations at Hell Gate. New York." *Minutes of Proceedings of the Institution of Civil Engineers*, 1886, pp. 264–274.

"The Harlem Ship Canal." *Harper's Weekly*, June 29, 1895, p. 605.

Harrison, John F. (Cartographer), et al. *Map of the City of New York: Extending Northward to Fiftieth St*. Published–New York: M. Dripps; Engraved and printed–Philadelphia: Kollner's Lithographic Establishment, 1852. Retrieved from the Library of Congress, www.loc.gov/item/2017586293.

Healy, Ryan. "The Strange History of NYC's Mighty Hell Gate." February 22, 2016. https://gothamist.com/news/the-strange-history-of-nycs-mighty-hell-gaten.

Heap, D P. *The Engineer Department, U.S. Army, at the International Exhibition, 1876*. Washington, D.C.: U.S. Government Printing Office, 1884.

"Hell Gate." *The American Universal Cyclopaedia: A Complete Library of Knowledge. A Reprint of the Last Edinburgh and London Ed. of Chambers's Encyclopaedia*. New York: S.W. Green's Son, 1882.

"Hell Gate." *Engineering News*, September 30, 1876, pp. 313–315.

"Hell Gate Ajar." *Indianapolis News*, September 25, 1876.

"Hell Gate Blown Up." *Sacramento Daily Union*, September 25, 1876.

"Hell Gate Destroyed—The Explosion." *Harpers Weekly*, October 14, 1876, p. 17.

"The Hell Gate Explosion." *The Popular Science Monthly*, November 1876, pp. 105–106.

"The Hell Gate Improvements." *Engineering and Mining Journal*, October 24, 1885, pp. 288–290.

"Hell-Gate, in the Channel of the East River." *Potter's American Monthly*, December 1875.

"Hell Gate; M. Maillefert's Blasting—His Work to Be Surveyed by the Government—Hallett's Point." *New York Times*, November 30, 1869.

"Hell Gate Open. Successful Destruction of Hallett's Point Reef. A Triumph of Science." *New York Herald*, September 25, 1876.

"Hell Gate—The Growth of the City." *New York Herald*, September 10, 1876.

"Hell Gate. The Present Channel—the Currents, Eddies and Rocks." *New York Herald*, January 30, 1870.

"Hell Gate the Scene of the Explosion." *Columbus Daily Enquirer*, September 30, 1876.

"Hell Gateless." *Rutland Weekly Herald*, September 28, 1876.

"The High Explosives in War." *The United States Army and Navy Journal and Gazette of the Regular and Volunteer Forces*. July 5, 1884, p. 993.

*Hurlgate and the Proposed Canal*. New York: Snosden [Snowden], [1832?].

"Improvements in New York Harbor." *Scientific Weekly*, August 27, 1881, p. 134.

"Inaugural Address." *Transactions of the Society of Engineers*, 1887.

Ingersoll, Charles Jared. *History of the Second War Between the United States of America and Great Britain, Declared by Act of Congress, the 18th of June, 1812, and Concluded by Peace, the 15th of February, 1815*. Philadelphia: Lippincott, Grambo, 1852.

Irving, Washington. *A History of New York: From the Beginning of The World to the End of the Dutch Dynasty, Containing, Among Many Surprising and Curious Matters, the Unutterable Ponderings of Walter the Doubter, the Disastrous Projects of William the Testy, and the Chivalric Achievements of Peter the Headstrong—The Three Dutch Governors of New Amsterdam: Being the Only Authentic History of the Times that Ever Hath Been or Ever Will*. London: W.C. Wright, 1825.

Irving, Washington. *Tales of a Traveller*. Part IV. New York: Library of America, 1991.

Jessen, Klark. "Remembering the Fall River Line." *MassDOT*, February 28, 2015. https://blog.mass.gov/transportation/mbta/remembering-the-fall-river-line.

"John Newton." *Prabook*. https://prabook.com/web/john.newton/1091924.

*Journal of the Senate of the State of New York*. New York: S. Southwick, 1850, p. 172.

Kellner, Arthur D. *New York Harbor: A Geographical and Historical Survey*. Jefferson, NC: McFarland, 2006.

Kent, James. *A Practical Treatise on Commercial and Maritime Law. With a Chapter on Incorporeal Hereditaments, etc.* n.p, 1837.

Klawonn, Marion J. *Cradle of the Corps: A History of the New York District, U.S. Army Corps of Engineers, 1775–1975*. New York: Department of Defense, Department of the Army, Corps of Engineers, New York District, 1977.

Klein, Daniel B., and John Majewski. "Turnpikes and Toll Roads in Nineteenth-Century America." eh.net/encyclopedia/turnpikes-and-toll-roads-in-nineteenth-century-america/.

Knoblock, Glenn A. *The American Clipper Ship, 1845–1920: A Comprehensive History*. Jefferson, NC: McFarland, 2004.

Knox, Thomas W. *Underground, or Life Below the Surface: Incidents and Accidents Beyond the Light of Day; Startling Adventures in All Parts of the World; Mines and the Mode of Working Them; Under-Currents of Society; Gambling and Its Horrors; Caverns and Their Mysteries; The Dark Ways of Wickedness*. Hartford, CT: J. B. Burr, 1874.

Kornblum, William. *At Sea in the City: New York from the Water's Edge*. Chapel Hill, NC: Algonquin Books, 2002.

Kouwenhoven, John A. *Columbia Historical Portrait of New York*. New York: Harper & Row, 1972.

Kroessle, Jeffrey A. *New York Year by Year: A Chronology of the Great Metropolis*. New York: NYU Press, 2002.

Lightfoot, Frederick S., and Harry Johnson. *Maritime New York in Nineteenth-Century Photographs*. New York: Dover Publications, 1980.

Lobell, Jarrett A. "The Hidden History of New York's Harbor." *Archeology Magazine*, November/December 2010. https://archive.archaeology.org/1011/etc/wtc.html.

Logel, Jon Scott. *Designing Gotham: West Point Engineers and the Rise of Modern New York, 1817–1898*. Baton Rouge: Louisiana State University Press, 2016.

"M. Maillefert's Blasting—His Work to Be Surveyed by the Government—Hallett's

Point." *New York Times*, November 30, 1869.

"Major General John Newton U.S.A." *Engineering News-Record and Contract Journal*, September 25, 1886.

Malcolm, Johnson. *Crime on the Waterfront*. New York: Penguin, 2005.

Marin y León, Juan J., and Addison Weed. *Destruction of the Reef at Hallet's Point*. Ithaca, NY: Cornell University, 1876.

*Marine Engineer and Naval Architect*, November 1, 1885, p. 197–198.

"Mary Newton at the Hell Gate Explosion." *The Pacific Rural Press and California Farmer*, November 4, 1876.

Matteson, George. *Tugboats of New York: An Illustrated History*. New York: NYU Press, 2007.

McKay, Richard. *South Street: A Maritime History of New York*. Brooklyn, NY: Going Coastal, 2004.

*The Military Engineer: Journal of the Society of American Military*. Alexandria, VA: Society of American Military Engineers, 1971.

Miller, William H. *Great Ships in New York Harbor: 175 Historic Photographs. 1935–2005*. Garden City, NY: Dover Publications, 2012.

Mines, John Flavel. *A Tour Around New York, and My Summer Being the Recreations of Mr. Felix Oldboy*. New York: Harper & Bros., 1892.

"Mining Operations at Hallet's Point." *Harper's Weekly*, September 23, 1871, p. 887.

"Mission." *National Academy of Sciences*. www.nasonline.org/about-nas/mission/.

*The Missionary Review of the World*, July 1889, pp. 547–548.

Mitchell, Henry. *Report on the Tides and Currents of Hell Gate: One of the Entrances to New York Harbor Made to the Superintendent of the Coast Survey*. Washington, D.C.: U.S. Government Printing Office, 1869.

"Modern Explosives and Their Practical Application." *Engineering News*, July 5, 1884, pp. 1–3.

Morgan, Christopher, and Edmund B. O'Callaghan. *The Documentary History of the State of New York*. New York (State) Secretary's Office. New York: Weed, Parsons, 1850.

Morrison, James L. *The Best School, West Point. 1833–1866*. Kent, OH: Kent State University Press, 1998.

Morrison, John Harrison. *History of New York Ship Yards*. Port Washington, NY: I.J. Friedman, Reissue edition, 1970; original, 1909.

Mowbray, George M. *Tri-Nitro-Glycerin, As Applied in The Hoosac Tunnel, Submarine Blasting, etc.* North Adams, MA: Robinson, 1872.

Munroe, Kirk. "Blowing Up Flood Rock." *Harper's Weekly*, October 17, 1885, p. 679.

National Academy of Sciences. "Biographical Memoir of General John Newton." *Biographical Memoirs*. Vol IX. Washington, D.C. National Academies Press, 1901, pp. 235–240.

"Navigating the East River." *Blue Seas*. https://www.offshoreblue.com/cruising/east-river.php.

"Nearing the End. The Great Blast at Hallett's Point on Sunday Next. General Newton's Proclamation." *New York Times*, September 17, 1876.

*The New American Cyclopaedia: A Popular Dictionary of General Knowledge*. Vol XII. D. New York: Appleton, 1873.

*The New Encyclopedia Britannica: Macropaedia: Knowledge in Depth*. Edinburgh: Encyclopedia Britannica, 2010.

New York City Government. "Transforming the East River Waterfront." https://www1.nyc.gov/assets/planning/download/pdf/plans-studies/east-river-waterfront/east_river_waterfront_book.pdf.

Newton, John. "East River and Hell Gate Improvements." *The Popular Science Monthly*, February 1886, pp. 433–448.

Newton, John. *Report of the Operations of the U. S. Steam-Drilling-Scow in the Removal of Way's Reef*. Washington, D.C.: U.S. Government Printing Office, 1875.

"Newton's Channel. Soundings and Operations at the Site of the Blast. The Reef Destroyed." *New York Herald*, September 26, 1876.

Nichols, Michael. *Hell Gate: A Nexus of New York City's East River*. Albany, NY: State University of New York Press, 2018.

Nichols, Theodore A. *The Caribbean Gateway to Colombia: Cartagena, Santa Marta, and Barranquilla and Their Connections with the Interior, 1820–1940*. Berkeley: University of California Press, 1951, p. 86.

"Nitro Glycerine Explosion." *New York Herald*, August 28, 1885.

"Nitro-Glycerine as Freight—An Important

Suit and Heavy Damages." *New York Times*, August 22, 1867.

"Nitroglycerine Its Use and Its Danger." *Scientific American*, February 1, 1868, pp. 73–74.

"Obituary." *Engineering and Mining Journal*, May 4, 1895, p. 419.

*Official Gazette of the United States Patent Office*. Washington, D.C: U.S. Department of Commerce, 1874.

"Opening Up Hell Gate." *The Pall Mall Budget: Being a Weekly Collection of Articles Printed in the Pall Mall Gazette from Day to Day, with a Summary of News*, October 22, 1885. pp. 9–10.

Paulding, James Kirke. *The Old Continental; or, the Price of Liberty*. New York: Paine and Burgess, 1846.

Perkins, Sid. "When Horses Really Walked on Water." *The Chronicle of the Horse*, May 21, 1999, pp. 90–92.

Pollara, Gina, et al. *The New York Waterfront: Evolution and Building Culture of the Port and Harbor*. New York: Monacelli Press, 1997.

"Preparing to Open Hell Gate." *Harper's Weekly*, October 10, 1876, pp. 660–662.

*Proceedings of the Board of Councilmen of the City of New York*. New York: The Board, 1862, p. 24.

Rattray, Jeannette (Edwards) *Perils of the Port of New York: Maritime Disasters from Sandy Hook to Execution Rocks*. New York: Dodd, Mead, 1973.

"Ready to Blow Up Flood Rock. Expected Effects of the Explosion." *New-York Tribune*, October 1, 1885.

"The Removal of Flood Rock, New York, by Tunneling." *Scientific American*, September 1881, n.p.

"Removal of Flood Rock, New York City." *Scientific American*, September 24, 1881, p. 197.

"Removal of the Hell Gate Rocks." *Scientific American*, September 30, 1876, pp. 214–216.

"Rending Hell-Gate Rocks; The Submarine Mine Exploded." *New York Times*, September 25, 1876.

*Report of the Board on Behalf of United States Executive Departments at the International Exhibition, Philadelphia, PA, 1876*. Washington, D.C.: U.S. Government Printing Office, 1884.

*Report of the Chief of Engineers*. Washington, D.C.: U.S. Government Printing Office, 1869.

*Report of the Chief of Engineers*. Washington, D.C.: U.S. Government Printing Office, 1870.

*Report of the Chief of Engineers*. Washington, D.C.: U.S. Government Printing Office, 1872.

*Report of the Chief of Engineers*. Washington, D.C.: U.S. Government Printing Office, 1875.

*Report of the Chief of Engineers*. Washington, D.C.: U.S. Government Printing Office, 1877.

*Report of the Chief of Engineers*. Washington, D.C.: U.S. Government Printing Office, 1879.

*Report of the Chief of Engineers*. Washington, D.C.: U.S. Government Printing Office, 1883.

*Report of the Chief of Engineers*. Washington, D.C.: U.S. Government Printing Office, 1886.

*Report of the Superintendent of the Coast Survey, Showing the Progress of the Survey During the Year*. New York: Robert Armstrong, public printer, 1869.

*Report on the Tides and Currents of Hell Gate: One of the Entrances to New York Harbor Made to the Superintendent of the Coast Survey*. Washington, D.C.: U.S. Government Printing Office, 1869.

Rodrigue, Jean-Paul. "The Emergence of Mechanized Transportation Systems." https://transportgeography.org/?page_id=995.

Rush, Thomas Edward. *Port of New York*. New York: Doubleday, Page, 1920.

Ruttenber, E. M. "Indian Geographical Names." *Proceedings of the New York State Historical Association. The Seventh Annual Meeting. With Constitution, By-Laws, and List of Members*. Hartford, CT: Case, Lockwood & Brainard, 1870.

Sanger, George P. *Public Laws of the United States of America. Carefully Collated with the Originals at Washington*. New York: Little, Brown, 1867.

Schermerhorn, Louis Younglove, et al. *Analytical and Topical Index to the Reports of the Chief of Engineers and the Officers of the Corps of Engineers, United States Army, Upon Works and Surveys for River and Harbor Improvement, 1866-[1892]*. Washington, D.C.: U.S. Government Printing Office, 1881.

Sears, Stephen. *Chancellorsville*. New York: Houghton, Mifflin Harcourt, 2014.

Sheard, Bradley. *Lost Voyages: Two Centuries of Shipwrecks in the Approaches to New York.* Locust Valley, NY: Aqua Quest, 1998.

"The Shock on the Bridge." *Brooklyn Daily Eagle,* October 10, 1885.

"Sketch of General John Newton." *The Popular Science Monthly,* October 1886, pp. 834–840.

"A Sketch of Pot Rock Hurl-Gate Channel." *New York Municipal Gazette*—Volume 1, issues 41–63 undated, p. 2.

Smith, R. A. C. *The Commerce and Other Business of the Waterways of the State of New York: A Tabulation of Facts About Waterborne Trade.* New York (State): Commission to Investigate Port Conditions and Pier Extensions in New York Harbor, 1914.

"State Tonnage Duties on National Commerce." *New York Times,* June 8, 1860.

Stevens, John Austin. "Progress of New York in a Century, 1776–1876. An Address Delivered Before the New York Historical Society, December 7, 1875." New York Historical Society, 1876.

Stone, William Lette. *History of New York City from the Discovery to the Present Day.* New York: E. Cleave, 1868.

"Story of the Man Who Conquered Hell Gate." *Greencastle Herald,* December 29, 1911.

Stoughton, Charles. "The Life Blood of a Nation Is in Its Commerce." To the United States Congress Assembled. Washington, D.C., December 1, 1879. www.loc.gov/resource/rbpe.20605200.

Striedinger, Julius H. "On the Simultaneous Ignition of Thousands of Mines and the Most Advantageous Grouping of Fuses." *Engineering,* September 21, 1877, pp. 236–237.

Susan Hua. "Oyster and Oyster Reef Restoration in the East River." *Restoring New York City,* Columbia University, December 20, 2006. http://www.columbia.edu/itc/cerc/danoff-burg/RestoringNYC/RestoringNYC_EastRiver.html.

"Tales of Destruction...Frozen Nitroglycerin." https://www.logwell.com/tales/frozen_nitroglycerin.html.

"Telegraphic News. Hell Gate. The Great Explosion a Complete Success. No Accident of Any Kind." *Daily Albany Argus,* September 25, 1876.

"Tides and Water Levels." *National Oceanic and Atmospheric Administration.* https://oceanservice.noaa.gov/education/tutorial_tides/tides03_gravity.html.

"To Be Wed Today." *New York Herald,* June 18, 1895.

"Transportation Developments in the Early Republic." *Conner Prairie.* https://www.connerprairie.org/educate/indiana-history/travel-and-transportation.

Tyson, Neil deGrasse. "America's Science Legacy." *Science,* November 20, 2015.

"The Unbarring of Hell Gate" *Scribner's Monthly,* November 1871, pp. 33–53.

United States Army Corps of Engineers. "The Conquest of Hell Gate." https://www.nan.usace.army.mil/Portals/37/docs/history/hellgate.pdf.

"United States Coast and Geodetic Survey. Report of the Superintendent–1874." ftp://ftp.library.noaa.gov/docs.lib/htdocs/rescue/cgs/001_pdf/CSC-0023.PDF.

*United States Congressional Serial Set.* Washington, D.C.: U.S. Government Printing Office, 1850.

*United States Congressional Serial Set.* Washington, D.C.: U.S. Government Printing Office, 1867.

"The United States Drilling Scow, East River." *Scientific American,* July 19, 1879.

United States Military Academy Association of Graduates. John Newton "Annual Reunion." June 10, 1895, p. 111.

van der Donck, Adriaen. *A Description of New Netherlands.* Lincoln: University of Nebraska Press, reprint edition, 2010.

van Pelt, Daniel. *Leslie's History of the Greater New York.* New York, 1898.

Vernon-Harcourt, Leveson Francis. *Achievements in Engineering During the Last Half Century.* New York: Scribner's, 1891.

Vowell, Sarah. *Assassination Vacation.* New York: Simon & Schuster, 2005.

Wainwright, Alexander. "Approaches to New York." *Harper's Magazine,* July 1884.

Waldman, John. *Heartbeats in the Muck: The History, Sea Life, and Environment of New York Harbor.* Guilford, CT: The Lyons Press, 1999.

Waldman, John. "Public Parks, Recreational Access, and the Post-Industrial Harbor of New York." *Gotham Gazette,* 2000. https://www.gothamgazette.com/commentary/comm.12.shtml.

*War Department Exhibit: Corps of*

Engineers, Catalogue. United States, n.p, 1893.

"What is Heat." *Cool Cosmos*. coolcosmos. ipac.caltech.edu/cosmic_classroom/ light_lessons/thermal/heat.html.

Wheeler, David Everett. "The New York Harbor and the Improvements Necessary for Its Accommodation of Commerce and the Removal of the Dangers at Hell Gate: A Paper Before American Geographical and Statistical Society." May 15, 1856.

Whitney, Ralph. "The Unlucky Collins Line." *American Heritage*, February 1957. https://www.americanheritage.com/unlucky-collins-line.

Wicks, Hamilton S. "American Industries, No.2." *Scientific American*, January 18, 1879.

Williams, Edward. "Statistics of the City of New York." *The New York Annual Register*. New York: J. Leavitt, 1832.

Williams, Edwin. *The New-York Annual Register for 1830–1837, 184: Containing an Almanac; Civil and Judicial List; with Political, Statistical and Other Information Respecting the State of New-York and the United States*. New York: J. Leavitt, 1832.

Williams, J. David. *America Illustrated*. Boston: De Wolfe, Fiske, 1883, 2003.

Wingate, Charles Frederick, and Henry C. Meyer. *The Engineering Record, Building Record and Sanitary Engineer*. New York: McGraw Publishing, 1900.

"Work at Newton's Channel." *New York Times*, December 26, 1876.

WPA Writers Project. *Maritime History of New York*. Brooklyn, NY: Going Coastal, 2004.

# Index